HAPPY HOUR
with
EINSTEIN

Inspirational Hors d'oeuvres
and Intellectual Small Plates

Dr. Melissa Hughes

Thinking about Thinking | Learning about Learning

Copyright © 2016 by The Andrick Group, LLC

All rights reserved. Copyright fuels creative thought, promotes a more diverse expression of ideas, and creates a vibrant culture of learning. Thank you for complying with copyright laws by not participating in or encouraging piracy of these materials in violation of the author's rights. Purchase only authorized editions. No part of this book may be reproduced, scanned, or distributed in any printed or electronic form without permission.

Cover illustration and design by Bradley Keppler.

"The mind that opens up to a new idea never returns to its original size."

INSPIRATION

Several years ago, I attended Conscious Capitalism and had the great pleasure of hearing Tony Schwartz and Simon Sinek speak. I was so excited about what I had learned, I shared it with total strangers... in the airport, at Starbucks, anyone who would listen. (Yeah... I was *that* lady!) That learning ignited a passion to not only continue on my quest to learn more about the brain, but also to share that learning with others. I had just begun using sketch notes as a learning tool, but what I learned from them is indelibly etched into my brain. I look back at these notes today, and I can remember the exhilaration I felt in the recognition that my brain was forever changed. I'm grateful...

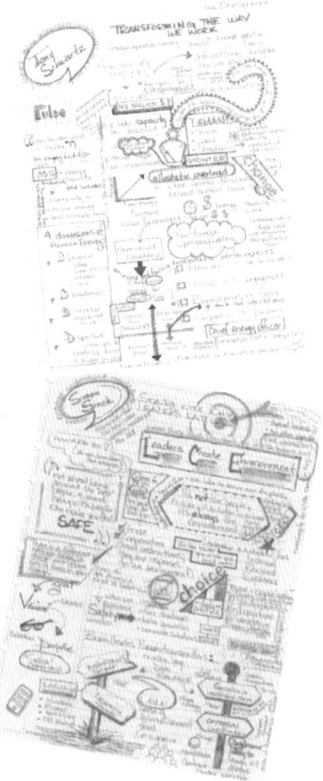

I'm grateful to Tony Schwartz for teaching me that pedaling harder and longer doesn't always get us there faster. We perform at our best when we move between spending energy and renewing our four core energy needs: physical, mental, emotional, and spiritual. You have changed the way I work and the way I think about work. [www.theenergyproject.com]

I'm grateful to Simon Sinek for demonstrating the power of being both *inspired* and *inspiring*. You have fueled my passion for thinking about thinking and learning about learning. Being so excited about learning something new that you can't wait to share it with others is perhaps one of the most powerful motivators on the planet. [www.startwithwhy.com]

Contents

Introduction 1

Part One

Einstein's Wild Ride 9
Think and Learn Like a Boss 21
Neuroscience 101: The Crash Course 31
The Cognitive Power of Sketch Notes 51
Laughter is the Best (Brain) Medicine 63
Get Your Groove On 73
Surprisology 83
The Grateful Brain 91
Happiness is the Secret to Success 103

Part Two

The Art and Science of Journaling 119

"All that is valuable in human society depends upon the opportunity for development accorded the individual."

Introduction

It was October 31, 2015. Baseball fans across the country were gearing up for Game 4 of the World Series. I was one of them. But I wouldn't be watching it on TV; I had a ticket to Citi Field. As luck would have it, the benefactor of this golden ticket lived in Princeton, so the weekend would begin in the Ivy League home of scientists, scholars, and scribes – and the "pope of physics," Albert Einstein.

As we meandered through campus on that crisp autumn day, I realized that the aura of the genius who died 60 years ago lives on. From the little single family house at 112 Mercer Street to the Institute for Advanced Study located at 1 Einstein Drive where he spent the last 20 years of his life, his presence is unmistakable. A bust on top of a 6-foot high granite pedestal at EMC Square memorializes Princeton's adopted son. But beyond the markers, plaques, and historical artifacts, you can almost feel him shuffling along the sidewalk next to you in that shaggy old sweater with a pipe in one hand and a stack of papers that would change the world in the other.

We made our way to The Alchemist & Barrister, a cool little pub nestled into a small side street with a few other bars and restaurants....maybe a place where Albert would have enjoyed a pint after a long day of quantum physics. What musings would he have shared? What stories would he have told? If you visit this local gathering place today, the first thing you'll see as you walk in the door is a large painting of Albert Einstein sitting at the bar… enjoying a pint.

(Printed with permission from artist, Morris Docktor, and The Alchemist & Barrister, 28 Witherspoon Street, Princeton, NJ.)

Albert Einstein died in 1955. Considering what we knew about the brain back in 1955 compared to today, it's hard even to contemplate that his contributions to society could have been greater. But, what if Einstein had the same depth and breadth of knowledge about cognitive function as we have today? What if he knew about neurotransmitters and other factors that impact our ability to learn? What if he had access to today's technology? It's almost impossible to imagine what he could have accomplished!

We once believed that intelligence and creativity were fixed. It was accepted that we were born with a certain capacity to learn, and we did the best we could with what we had. Now, we know differently. Since the 1990s, also known as "the decade of the brain," we've discovered that we can actually increase our capacity for learning, problem solving, creativity, success, and happiness. Technological advances and a wealth of research have proven that innovation, critical thinking, and problem-solving are hardwired into each of us. We all have the same anatomy between our ears. The difference between the best and the rest may lie in our understanding of the factors that impact how the brain functions.

Perhaps some of us just have a better understanding of the brain than others. When was the last time you thought about how you learn? With a basic understanding of how the brain works and by applying a few simple brain-based strategies, we can all increase our learning power. Learning how to learn – or relearning with the whole brain – is the first step.

My roots are deep in education. I spent many years teaching students from elementary school to the university level, and my professional growth was grounded in the premise that I couldn't master my craft of teaching unless I had a solid understanding of how the brain works. The College of Education at the University of Akron filled my cognitive backpack with instructional methods and educational theories that qualified me to teach children what they needed to know to be labeled "proficient" and move on to the next grade. But, I always found it curious that in all those years of preparation, no one ever taught me how the brain works – how we actually learn and what impacts our capacity for intelligence, problem-solving and creativity. I was a serious student, and I could tell you everything about Piaget's constructivist theory of knowing or Gardner's multiple intelligences. Unfortunately, none of those courses included how cortisol actually impedes the brain's ability to function properly or that laughter releases dopamine and an array of endorphins that enhance cognition. Education is the only profession in which building the capacity to learn is the primary task – every single day – and I had never been taught how the brain actually works!

Since those early days in the classroom, my journey has been guided by the desire to discover more about cognitive function and maximizing the brain's efficiency. Given that we all have the same pieces and parts above the shoulders, what makes some people incredibly talented, creative, intelligent, or innovative? What is it that makes some people super productive while others lack motivation? Why do some people avoid challenges while others embrace them?

Whether you're five years old or 50 years old, the ability to learn is a skill that each of us can develop, improve and grow. With some basic knowledge and a few minutes of intentional daily practice, we can expand that capacity to learn. This book is a collection of recent findings of the human brain that you can apply in your own life for greater productivity, creativity, success, and happiness.

Part One is a "crash course" on neuroscience. This section is intended to give you a basic understanding brain structure and cognitive function. The first few chapters explain the basic architecture of the brain and identify factors that impact our capacity for intelligence and learning focusing on three key concepts: neurogenesis, neurotransmitters, and neuroplasticity. Each subsequent chapter drills down into a specific behavioral aspect of whole-brain engagement and explores the impact each has on cognition, creativity, and well-being. The "Reflect and Apply" exercises at the end of these chapters provide the framework for individual application and deliberate practice. Don't skip these pages! Studies show that when we recall information and process it through a personal lens, we're better able to apply that information in a meaningful way. Not only will these pages help you incorporate this information into your own life, but they also serve as an excellent basis for group discussion and organizational learning.

Part Two gives you a simple strategy to apply this new knowledge with intentional focus in just a few minutes each day. The 28-day journal guides you through the process of applying neuroscience to your work, your relationships, and your life. The journal pages may look pretty, but the format is more than just aesthetics. Studies indicate that free flowing shapes, textures, natural colors and abstract images can help us break free from the rigidity of everyday life, enable us to make emotional and cognitive associations, and activate brain states that are otherwise harder to access. Consider it a four-week neuroaesthetic experiment with the potential to change the way you approach problems, set goals, and interact more effectively with the people in your corner of the world.

As we continue to explore the mysteries of the human brain, the very nature of science pushes us to replicate, scrutinize, discover flaws and test new theories. Undoubtedly, some of the studies presented here may be challenged with critical examination, additional research, and opposing arguments. The intention is not to substantiate or discredit any particular study, nor is it an exhaustive look at the body of research. This book won't get you a degree in neuroscience, and don't expect the content and style to be that of a scientific paper or medical symposium. Think of it more as happy hour with Einstein and a few other brainiacs sitting around the bar sharing recent brain-based research in plain speak over a few cocktails.

Let the journey begin!

Cheers!

Part One

"There are only two ways to live your life.
One is as though nothing is a miracle.
The other is as though everything
is a miracle."

Chapter One
****
Albert Einstein's Wild Ride

It's inevitable that the journey to understanding brain function and intellectual capacity would begin with Albert Einstein. His intellectual achievements have made his name synonymous with the word *genius*. What enabled a man who was described by his earliest teachers as lazy, sloppy and even "mentally retarded" to conceptualize space, time, mass, and energy in a way that would literally change the world? How was his brain so different? Thanks to the pathologist who secretly smuggled his brain from the autopsy room, scientists have been exploring that question since April 17, 1955.

What would unfold into a bizarre journey for Einstein's brain began with severe chest pains on that fateful evening. He was admitted to Princeton Hospital and died the next day of an aortic aneurysm. His last will and testament unequivocally instructed his body to be cremated with the ashes scattered secretly, but an autopsy would be conducted.

Dr. Thomas Harvey wasn't even scheduled to perform the autopsy and prepare the body for cremation. He was the "on call" pathologist that night and a last-minute replacement. The young Dr. Harvey found himself alone in the morgue looking down at one of the most amazing scientists in history. At that moment, he decided that the world could not - *would not* – be finished learning from the genius.

Dr. Harvey left the hospital that day carrying a duffle bag that contained Einstein's brain as well as his eyeballs which he later gave to Henry Abrams, Einstein's ophthalmologist. Abrams ultimately placed them in a safety deposit box in New York City (where they remain to this day). Princeton Hospital urged Harvey to turn over the brain, but he refused. A few months later, he was fired. We can debate all day long about whether Harvey was a hero, opportunist, or criminal, but Einstein's son eventually granted him permission to study it. In the name of science, Albert's wild ride began.

Harvey was not a brain specialist, and at some point, he realized that he didn't have the skills or the expertise to conduct the study. Harvey sliced Einstein's brain into 240 pieces. He sent small slivers to a handful of the best and brightest neuroscientists around the world and waited for them to report back with their findings. Meanwhile, he guarded the remainder of Einstein's brain as the most significant treasure of the 20th century. As he waited for these hand-picked scientists to unlock the secrets of this great mind, Harvey was both secretive and protective of what would be, according to him, a contribution to science on par with those from Einstein himself.

For the next 20 years, we didn't hear much about either the brain or the research. In 1978, 27-year old journalist, Steven Levy would put both back in the headlines. As it turns out, Levy's editor was fascinated with Einstein and tasked his young reporter with finding out what happened to his brain. Levy's old-school investigative reporting started with a call to 411 which ultimately led him to a Dr. Thomas S. Harvey located in Wichita, Kansas. Levy simply called him up, and the two agreed to meet on a Saturday afternoon in Harvey's little office. Reluctantly, Harvey began to share his story. Then, perhaps overtaken by pride, he revealed that the brain was right there – in that tiny little office!

Behind a Styrofoam beer cooler, inside an old Costa Cider box, and buried under crumpled newspapers was a mason jar that contained gray, spongy chunks of human tissue. Harvey confirmed the contents of the jar were the remains of Einstein's cerebellum, cerebral cortex, and aortic vessels.

After Levy's article was printed later that summer, Harvey's quiet life would be interrupted with a flock of reporters who camped out on his lawn as well as scientists who wanted in on the research. He became the top coach in the *International Neuroscience League* as he determined who would play on "Team Einstein."

Dr. Marian Diamond was one of his top draft picks. Harvey cleaned out an empty Kraft Miracle Whip jar and sent four small pieces to the Berkeley neuroscientist by U.S.P.S. By 1984, Dr. Diamond was ready to announce her important discovery to the world. Dr. Einstein had more glial cells than the average brain – especially in the association cortex (the part of the brain responsible for complex thinking and imagination). For a long time, neuroscientists thought that the glial cells were just the brain's maintenance and housekeeping staff – delivering nutrients, repairing neurons, and clearing out all of the dead cells on their way out.

Additional research, however, established that while the neurons get all of the attention for brain activity, glial cells are actually responsible for synapse function. Those tiny gaps between neurons use synapses to communicate process information. We can thank our glial cells for the information highway system that moves data faster than a Formula I race car.

Well, now it was all beginning to make sense. Einstein had more glial cells which made his neurons function and communicate better. Obviously, this was big news! Diamond's work gained a lot of attention in the science community. Some celebrated the discovery while others poked holes in it. (Haters are gonna' hate.) It didn't take long before her colleagues began examining her work with intense scrutiny and criticism. Without boring you with all the scientific details, her groundbreaking discovery was ultimately exposed as critically flawed with selection bias.

Over the next few decades, a select number of respected scientists would have the opportunity to look at Dr. Einstein's brain. We'd learn that his frontal cortex was thinner than average but more neuron dense. We also learned that Einstein had an extra ridge on his mid-frontal lobe, the part used for complex planning and working memory.

In 1999, Canadian neuroscientist Sandra Witelson would publish a paper entitled *"The Exceptional Brain of Albert Einstein."* In it, Witelson claimed that while Einstein had more glial cells and neurons, the real difference was in the fissure patterns. Albert didn't have much of a lateral fissure. Also called the Sylvian fissure, this major furrow separates the parietal lobe and the frontal lobe. Since the parietal lobe handles mathematical ability, spatial reasoning, and three-dimensional visualization and the frontal lobe is responsible for executive functioning, this seemed pretty significant for the guy who envisioned a ride through space on a beam of light and translated it into the theory of relativity.

After safeguarding the brain for over 40 years, Harvey returned to Princeton. However, his odd obsession with Einstein wasn't over. Then 84-year old Harvey and a freelance writer named Michael Paterniti set out on a cross-country road trip to meet Al's granddaughter,

Evelyn Einstein. Like some weird spinoff of *Weekend at Bernie's*, Albert's brain sloshed around in a Tupperware container stowed in the trunk of their rented Buick Skylark from New Jersey to California. Paterniti chronicles the bizarre excursion in his book, *Driving Mr. Albert: A Trip Across America with Einstein's Brain*. Evelyn didn't want the brain, so the duo popped the brain back in the trunk and returned to Princeton where they would part ways. A year later, fueled by a sense of either guilt or responsibility, Harvey quietly returned Albert's brain to Princeton.

Fast forward to 2013. A team of scientists led by Weiwei Men at East China Normal University's Department of Physics made the most remarkable discovery yet. It turns out that beyond his abundance of glial cells and neuron-dense frontal cortex, he had a freakishly large corpus callosum. The biggest nerve fiber bundle in the brain connecting the two hemispheres was thicker and larger than normal. An undersized lateral fissure combined with an oversized corpus callosum meant that Einstein's brain was more well-connected than most. He was able to think, learn, and explore the world around him with his whole brain.

In the years leading up to this discovery, we had already accumulated lots of evidence suggesting the right brain/left brain theory was an overly simplified explanation of cognitive function. The most recent studies suggested that deeper cognition results from the dynamic interactions of brain regions operating in large-scale networks. We'll explore all of this further in the next chapter.

Could it be that the difference between the best and the rest lies in the size and efficiency of those networks? If we can learn how to improve the networks that enable the regions of the brain to communicate, perhaps we can create the conditions necessary for deeper cognition and more enlightened understanding. The key to understanding the way the brain processes information lies not only in the knowledge of neural networks but also in recognizing that different patterns of neural activity are important at different stages of the cognitive process. This is the essence of whole-brain thinking and learning.

Einstein would probably have been very proud to know that he's continued to make significant contributions to science long after his death. As for Thomas Harvey, he never realized the full impact of his actions in the morgue at Princeton Hospital back in 1955. He died in 2007 at the age of 94.

The Autistic Savant

When Kim Peek was born in 1951, doctors found an abnormality on the right side of his skull, similar to hydrocephalus. Initially diagnosed as mentally retarded, Kim was later re-diagnosed as an "autistic savant." Before he was two years old, he was able to read and memorize picture books. One by one, he would read them, memorize them, and then return them to the shelf upside down to indicate he had finished them (a practice he continued throughout his lifetime). Even as a toddler, he could speed read vast amounts of information and accurately recall it all later. However, other developmental milestones didn't come quite as easily. For example, he didn't start walking until he was four years old, he had difficulty with ordinary gross motor skills, and he scored well below average on standard IQ tests. While he could remember countless books word for word, he couldn't remember where to find the silverware. He began to demonstrate socially unacceptable and repetitive actions such as pacing and rocking. By the time he was 6, his doctor diagnosed Kim as "hopelessly retarded" and recommended a lobotomy. His father, Francis refused. He was expelled from school at age 7 for an uncontrollable outburst that lasted 7 minutes, and was home-schooled by tutors until he was 14 when he completed the high school curriculum.

At age 18, he got a job as a payroll clerk for a mid-size company. He calculated payroll for 160 people without ever using a calculator. Several years later, his employer let him go and eliminated his position when they decided to computerize payroll. Kim then became a bit of a local celebrity appearing in schools and public venues astounding the crowds by performing complex mental calculations in his head or identifying the day of the week associated with any date in history.

His neuropsychiatrist began studying his brain via MRI images in 1988 and identified abnormalities such as damage to his cerebellum. But it wasn't until 2004 that neuroscientists at the Center for Bioinformatics Space Life Sciences discovered that his two hemispheres were not separated at all. Rather, they created a single large data storage area that enabling the neurons to make unusually fast connections. Scientists theorized that this explained how he could read and recall the contents, word for word, of more than 10,000 books. Kim simultaneously read the left side of a page with one eye and the right side with the other, and dumped all of the data into a massive memory bank. Because there was no separation of the two hemispheres, he was using his whole brain unlike the rest of us.

Kim Peek died of a heart attack in 2009, but not before he met Dustin Hoffman – the actor who would eventually snag an Academy Award for playing Peek as Raymond Babbitt in the movie *Rain Man*.

Get Your Whole Brain in the Game

Few have the cognitive capabilities of Albert Einstein or Kim Peek. And while they seem to be on opposite ends of the neuroscience spectrum, they are both geniuses who, among other differences, used the whole brain to process information. There are a few simple ways to engage the whole brain and strengthen your corpus callosum. Brain scans show that playing an instrument, for example, not only uses more of the brain, but also promotes brain plasticity and neurogenesis. (By the way, Einstein was also a gifted violinist and pianist.)

Music and brain function are inextricably linked. The impact of music on complex perception, cognition, memory, and motor function has been studied since 1933. The landmark "Mozart Effect" study raised the controversial claim that listening to a Mozart sonata for ten minutes would increase one's IQ has been widely criticized, but it did open the door for significant research proving of the effects of music on the human body, brain function, and brain cell regeneration.

Over the last decade, neuroscientists have been using technology (fMRI, PET, EEG, MEG scans, etc.) to learn more about the impact of music on brain function. The characteristics that make up a given piece of music – wavelength, tone, hertz, timber, pitch, etc. – affect us in a variety of ways.

For example, rock music, especially songs with heavy bass, can infuse a sense of power-related thoughts and behavior. The clarity and elegance of classical music have been shown to improve focus, memory, and concentration. Slower Baroque music creates a mentally stimulating environment conducive to tap into higher cognitive tasks.

Impressionist music like Debussy and Ravel can stimulate the imagination and tap into your unconscious where many of your creative impulses live. Unfamiliar music triggers abstract thinking and helps generate creative ideas. Jazz and "new age" music with no dominant rhythm can also promote a sense of relaxed alertness and inspire creativity. Simply listening to the right type of background music while performing cognitive tasks is the brain's equivalent of a cross-training workout.

Since the left side of the brain controls the right side of the body and vice versa, activities that require you to coordinate both hands at the same time is a good way to stimulate and strengthen the corpus callosum. Juggling, kayaking, power walking or anything that puts both sides of the body in motion will put your whole brain in motion, too.

You can learn how to build ambidexterity with simple everyday activities. For example, draw simple shapes, practice writing your name or using the computer mouse with your non-dominant hand. Brush your teeth or use a fork with your non-dominant hand. Visualize what is happening in the brain as you do. It may feel awkward at first, but eventually, it will become more comfortable.

While neuroscientists continue to explore cells, synapses, fissures, and regions of the brain, we can put whole-brain engagement to use. Just knowing and applying the fact that greater brain engagement leads to deeper cognition, and greater creativity can make a big difference in the way we approach daily tasks.

Reflect and Apply

Knowing what you know now, how have your thoughts about the right-brain/left-brain theory and the whole-brain approach to thinking and learning changed?

Think about a time when you were "in the genius zone." What were the circumstances? What was the environment? What do you think was happening in your brain?

Chapter Two
****
Think and Learn Like a Boss

We once believed that people were either "right brained" or "left brained." The creative artists, musicians, and poets were right brained, and the analytical mathematicians, programmers, and engineers were left brained. Now we know that entire theory is an oversimplification. While specific regions of the brain are responsible for specific things, we are aware that the brain functions as a whole system. The better the regions integrate with each other, the better they all work independently. This construct of a "whole-brain thinking" has evolved into a sound framework for learning and improving cognitive performance.

Advances in technology opened the door to faster and more in-depth research than was possible when the "left brain-right brain" theory was accepted. Back then, the only way we could see what was happening in the brain was through invasive procedures which were largely done on people who suffered from mental health problems or neurological diseases. Today, we can actually see what happens in a healthy brain when exposed to virtually any kind of stimulus while sitting comfortably connected to a machine with a few wires and patches. There is a wealth of research that explores how the brain reacts to loud music, soft music, classical music, horror movies, angry faces, friendly faces, colors, chocolate, alcohol, drugs… the list goes on and on.

In the education world, these findings are changing the way educators deliver instruction. The traditional "sit and listen" environment has evolved to "maker-spaces" and other more engaging hands-on approaches to learning. We're seeing tremendous academic gains when students are exposed to a wide variety of fine and gross motor movement, rich visuals, collaborative projects, and a strong integration of the arts as core components of learning experiences. While students have individual learning styles and preferences (which may be right brain or left brain dominant), when they engage the whole brain they demonstrate deeper cognition, greater creativity, and improved problem-solving.

The good news is the benefits of whole-brain thinking and learning are not confined to the classroom. Engaging the whole brain enables learners and thinkers *of any age* to approach cognitive tasks, creative endeavors, and even triggers to stress more efficiently. Therein lies the key to competitive advantage in the workplace. All things being equal, those organizations that foster the collective knowledge, talents, skills, and creativity of their entire workforce will, quite literally, outsmart the competition.

The C-suite of the Brain

Think of the structure of the brain in terms of the Apple organization. The leaders on the executive team all have specific roles and responsibilities in place to keep over 110,000 employees doing their jobs. When that team is competent and working together cohesively, people are engaged, productive, and working efficiently.

The brain has a leadership team, too, and when they all work together, they enable the 100 billion neurons to keep the entire body operating at maximum efficiency. When that team effectively manages the mental, physical, emotional, and social performance, the whole body works better. Think of your "neurological headquarters" as being divided into four departments: left, right, upstairs and downstairs.

You already know that the brain is split into two hemispheres. Each side is not only anatomically different but also responsible for very different tasks. Your left brain loves order. It is logical, linear, and linguistic. It's the list-maker. Even the alliterative description of the left brain is aligned – all l-words! The left brain is where scientists and mathematicians hang out. Your right brain is creative, intuitive, and emotional. It deals with sensory, experiential feelings. This is where music, art, and poetry are created and enjoyed.

While we once classified people as either "right-brained" or "left-brained," we now know that it's much more complex than that. Specific brain functions take place in specific regions of the brain, but those regions coordinate with one another for data input, processing, and output. For example, a computer programmer may exercise the left side of his brain more than the right side due to the work he does, but that doesn't mean he is left-brained. He uses many regions of the brain to breathe, see, read, write, compute, etc. Dominance is dependent upon the task, not the person.

In addition to having two hemispheres, we also have an upstairs brain and a downstairs brain. The downstairs brain takes up space from the bridge of your nose down to your chin. It includes the brain stem and the limbic system, and this is where basic functions like breathing, blinking, balance, and other involuntary impulses take place. High emotions and instincts also happen downstairs. Scientists sometimes refer to this as the primitive brain, or the *reptilian brain*, because it takes care of basic needs that we do without conscious awareness.

The upstairs brain is much more complex and evolved. This is where more sophisticated mental activities such as planning, thinking, and executive functioning happen. This is also where we experience empathy, distinguish between right and wrong, and understand the consequences of our actions. With a fully functioning staircase, the primitive brain (emotional) and the sophisticated brain (intellectual) inform one another in every brain-related task and function.

Now that we have a basic understanding of the brain's C-suite let's drill down a bit deeper into the org chart. Although the minor wrinkles in each brain are unique (like fingerprints), our major wrinkles and folds form 4 different lobes in each hemisphere:

Frontal lobes
This is where we manage decision-making, problem-solving and thinking. The frontal lobes (including the prefrontal cortex) is the region often called the executive control center. This region monitors higher-order thinking, logic, emotional and rational control. Our personality (also called "the self-will" area) lives here, too.

Temporal lobes
The temporal lobes are located above the ears. They enable you to process sound, music, facial recognition, and speech. This region also helps with long-term memory and feelings.

Occipital lobes
Occipital lobes are at the back of your head. This is the central image processing center. Similar to the way the temporal lobe makes sense of auditory information, the occipital lobes help you process visual input so that you can understand what you see.

Parietal lobes
The parietal lobes process senses, spatial orientation, some types of recognition like colors and shapes, and a few specific academic skills like reading and math calculations.

The lobes all work together with a few other regions:

Motor cortex
The motor cortex is located between the frontal lobes and the parietal lobes and spans across the top of the brain from ear to ear. This little band works with your cerebellum to coordinate the learning of motor skills.

Limbic system
The limbic system is squished in the center of the brain right above the brain stem. This is the highly complex part of the brain that manages the interplay between emotion and reason. The limbic system is the conductor for managing learning, memory and emotional processing in the thalamus, hypothalamus, hippocampus, and amygdala. It is sometimes referred to as the "mammalian brain" because it adds a layer of emotion and control to the automatic responses of the reptilian brain.

Cerebellum
The cerebellum sits right behind the brain stem and comprises only about 10% of your brain's total weight. Yet, it contains more neurons than the rest of the brain regions combined. The cerebellum monitors all of the impulses sent from the nerves in the muscles and coordinates movement fine and gross motor movements. It calibrates the detailed form of

movements rather than decides what movements to execute. This is the part of your brain you should thank when you make that "hole in one." It also stores the memory of the certain movements we do without even thinking about such as tying your shoe or lifting a glass of wine to your lips.

Corpus Callosum
Technically, the corpus callosum is not recognized as one of the main brain regions. However, this broad bundle of neural fibers is what connects the two hemispheres and facilitates communication between them. It's hard to talk about whole-brain thinking and learning without giving props to the part that makes it all happen.

Bringing the Whole Team to the Grown-Up Table

Now that you know the main parts of the brain and the responsibilities of each, it's easier to understand the concept of whole-brain thinking and learning. The brain is highly specialized, and the degree of specialization determines how we think or what captures our attention. The prefrontal cortex is the boss. It controls all decision making, planning and problem solving. It's in charge of pretty much all the heavy lifting. Now that we can actually see inside the brain, we know that the boss can get the job done more efficiently and effectively when the cerebellum temporal lobe and occipital lobe show up to help. Imagine you're tasked with coming up with creative ideas for a new product launch. The prefrontal cortex is the department responsible for generating the messaging, the vision for graphics, and go-to-market strategy.

However, if the boss is smart, he'll invite the temporal lobe to the table (perhaps with a little soft background music by Claude Debussy). The specialized skills of the temporal lobe will enable the prefrontal cortex to tap into the unconscious where the creative impulses live. We also know that jazz and "new age" music with no dominant rhythm can promote a sense of relaxed alertness that helps inspire creativity and generate a new idea. Also, if the ideas are sketched out on a whiteboard or chart paper, now the cerebellum and occipital lobe are in the game, too. The construct of whole-brain thinking and learning simply means that more regions of the brain are activated which means there is more neural activity going on to tackle the task.

The key piece of whole-brain thinking is that the brain is designed to work as a system, not as individual parts. We do not function with "half a brain" as the terms "left brained" and "right brained" imply. At a time when creativity and innovation are as important as skills and experience, thinking with the "whole brain" pushes us to think in terms of *and* instead of *or*.

Remember, this book isn't designed to get you through med school. This is happy hour. So before you start showing off your neuro-knowledge and shouting out parts of the brain that weren't mentioned here, these are just the hors d'oeuvres and small plates. With 100 billion neurons firing 200 times per second, a lot is going on up there.

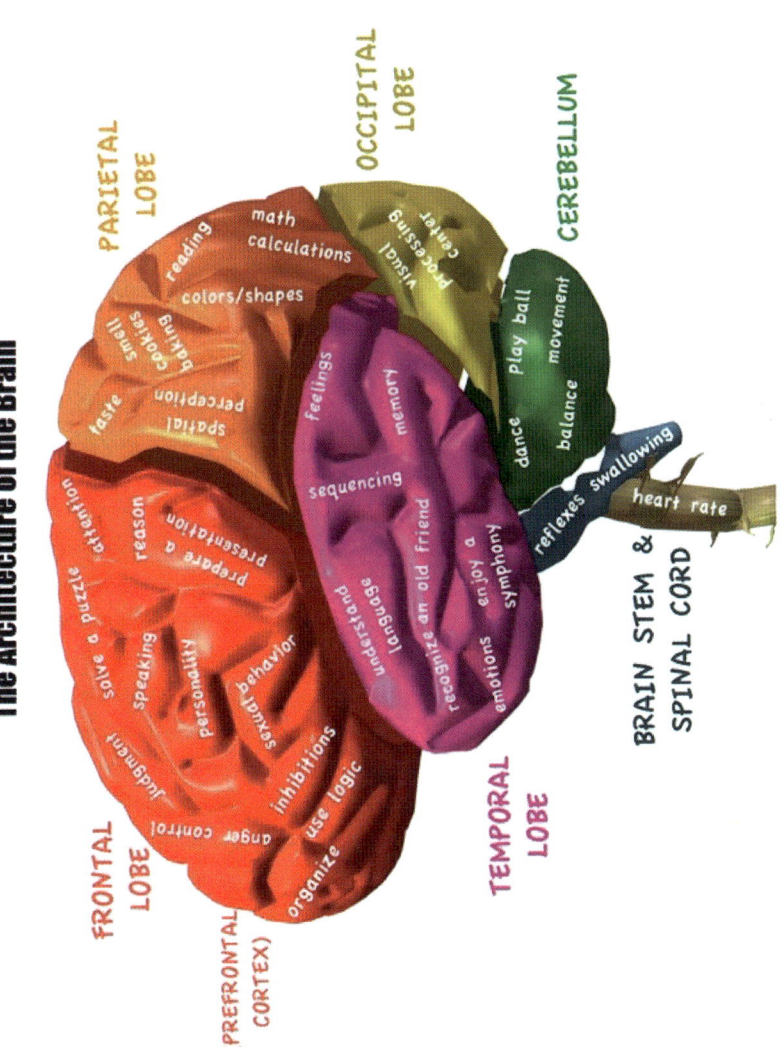

Reflect and Apply

Think about a typical day. What regions of your brain are the most active? What regions are the most inactive?

What percentage of your brain do you think you use?

The contention that we only use 10% of the brain is a myth. The human brain uses almost one-fourth of the body's total energy supply. While it's true that at any given moment not all regions are engaged and firing, most are continually active even during sleep. During the course of the day, you use 100% of the brain. To illustrate, consider what it takes to drink a glass of wine: reach for the bottle and remove the cork with the corkscrew, pour it into a class, swirl the wine, raise and tilt the glass to observe the legs, breathe in the aroma, and then finally bring the glass to your lips for that first taste. Even if you skipped a step or two, the occipital, frontal, and parietal lobes, sensory motor cortices and cerebellum are all engaged in the time it takes you to open the bottle, pour, and say "Cheers!"

Chapter Three
****

Neuroscience 101: A Crash Course

Did you know…..

95% of your decisions take place in your subconscious mind.

A piece of brain tissue the size of a grain of sand can contain up to 100,000 neurons all communicating with one another.

We used to think we only use 10% of the brain. We now know that we use most of our brain most of the time, even when we're sleeping.

The average person generates approximately 50,000 thoughts per day.

Information travels in the brain via electrochemical transmissions at a speed of up to 270 mph. Formula I race cars max out at 240 mph.

The average attention span of a human is 8 seconds. The average attention span of a goldfish is 9 seconds.

The human brain is so amazing it will blow your mind!

I'll be honest; I get more than a little geeked out over neuroscience. I unabashedly own that. To me, finding a shiny, new brain nugget is a little like unwrapping a present that I want to both savor and share. The more I learn about the brain and our ability to impact how it works, the more I want to share it with just about anyone who is the least bit interested in listening.

I realize that "Hey, let's grab a glass of wine and talk about the latest neuroscience discoveries!" isn't the best way to make or keep friends. Let's face it... the very word *neuroscience* can be a little intimidating to people. But in my conversations, I've found that most people share an innate curiosity about the spongy, 3-pound mass of tissue that controls every aspect of the body from our heart rate, immune system and emotions to our capacity for learning, memory, and creativity.

The field of neuroscience is still in its infancy, but the rapid explosion of knowledge over the last few decades has illuminated two significant things: (1) just how misinformed we once were and (2) how much we still have to learn. In simplest terms, neuroscience is the scientific study of the nervous system. While the scope of the field has grown over the past several decades to include the study of molecular, cellular, developmental, structural, functional, computational and medical aspects of the nervous system, this is a crash course focusing on three key concepts: neurogenesis, neurotransmitters, and neuroplasticity.

Neurogenesis

If you know the difference between the Charlie Brown Christmas tree and the Rockefeller Center tree, you're halfway to a basic understanding of neurogenesis. The term neurogenesis is made up of the word parts "neuro" meaning "relating to nerves" and "genesis" meaning "the formation" or "creation." Neurogenesis just means the formation and development of neurons.

The human brain has more than 100 billion neurons. The neuron is the basic working unit of the brain designed to transmit information electrochemically. Brain neurons begin forming about four weeks after conception. As the fetus continues to grow and experience stimuli, the neurons begin to grow dendrites and axons. The axons send electrochemical signals, and the dendrites receive them through small gaps at the ends of the dendrites called synapses.

Neural growth in the womb is extraordinary at a rate of 250,000 neurons per minute, and they all work overtime to create the dendrites, synapses, and axons necessary to communicate with one another. This process is called dendritic branching, and this is the foundation for all brain activity.

The process of dendritic branching continues at an incredible rate in the baby's first few years of life. At birth, the number of synapses per neuron is 2,500. By age 2, that number increases to more than 15,000 per neuron. This proliferation of neuron growth is called transient exuberance because it's temporary. The brain's natural pruning process kicks in to clean up and organize all of those unused, malformed, or misconnected dendrites. The process of growing and pruning dendrites continues throughout our lives (the "use it or lose it" principle applies here), but those first few years set the stage for healthy dendritic branching and cognitive growth throughout our lives.

After the neurological landscapers come through for that first major pruning, our dendrites look like the Charlie Brown Christmas tree. However, between the ages of 3 and 12, the brain continues to grow dendrites at an astonishing rate. Think about how a toddler begins to explore the world. Everything he sees and experiences is a discovery. Everything is novel. The more the child experiences, the more branches – or dendrites – grow as he learns sounds, shapes, colors, safety, fear, love, and a multitude of other things. Neurogenesis doesn't stop at a certain age. The more we learn, the more dendrites we grow. The more dendrites we have, the greater capacity we have to grow more dendrites and the connections necessary to communicate with each other. In time, all of that growing and learning and pruning enables the Charlie Brown Christmas tree to grow into the Rockefeller Center tree.

Neurotransmitters

The limbic system is the emotional control center of the brain. This is where we manage most of our emotions like happiness, sadness, fear, anger, and empathy. Emotion plays an enormous role in the brain's capacity to learn and function properly. The stronger your emotions are toward a specific event or experience, the stronger your limbic system responds to either impede or enhance cognition and file that experience away in long-term memory.

Located just beneath the cerebral cortex, the limbic system is directly responsible for releasing chemicals called neurotransmitters in response to certain emotions. The particular emotion you feel determines which neurotransmitter the limbic system produces. Cortisol, for example, is released in response to fear or stress as part of the brain's fight or flight mechanism. The brain produces other chemicals like serotonin, oxytocin, and endorphins when we feel positive emotions like pleasure, happiness, and love. Others such as dopamine, glutamate, and GABA (an amino acid which acts as a neurotransmitter by inhibiting nerve transmission) play important roles in cognition and brain circuitry.

These neurotransmitters help us regulate our mood and emotions and have an enormous impact on our performance and overall well-being. In addition, they play a major role in cognition, memory, and resilience. Put simply, they can significantly impede or enhance executive function, self-regulation, memory, and creativity.

Essentially, as neurotransmitters help us regulate our emotions, they also serve as the on/off switch for cognition and learning. If the limbic system is fighting stress, it stops everything else to conquer the threat. That's the way it's supposed to work. Imagine you're peacefully enjoying a pleasant summer day when, out of nowhere, a big, growling, black bear approaches.

Your sympathetic nervous system kicks into gear and activates the release of norepinephrine to prepare you for flight, fight, or freeze. Also known as an *amygdala hijack,* an army of chemical messengers signal the production of cortisol. Cortisol shuts down the rational brain by allocating all of the brain's resources to address the threat. Blood pressure and blood sugar levels increase to create a boost of energy. Your heart rate and breathing also increase which circulates the cortisol throughout the whole body. All of this happens without you even thinking about it to give you the increased strength or speed necessary for survival.

While few of us are faced with a growling bear, the same physiological changes occur when we experience stress. The intensity of the emotion brought on by the stimulus determines the magnitude of the physiological response. Common stressors such as a difficult boss or coworker, financial constraints, or even just being stuck in traffic can hijack the amygdala and put the rational brain on lock-down.

The Science of Stress

The two emotions that are specifically associated with stress are fear and anger. Anger is an energizing emotion, and fear is an energy-draining emotion. Both are manifestations of stress. None of us are immune from stress. It doesn't feel good; it makes us irritable, grumpy, distracted and tense. But, it's impossible to eliminate stress completely. A certain amount of stress – and the right kind – is actually a good thing. The key is not letting the good stress take over and become the bad stress.

There are two main types of stress - acute stress and chronic stress. We need a certain amount of acute stress to keep the brain active and primed for high performance. Rushing to meet a deadline, the loss of an important contract, or getting a speeding ticket are all examples of acute stress. A healthy brain kicks into alert mode to process the event and eventually shift back to a sense of calm when the danger or stress has been managed.

Studies conducted by Daniela Kaufer and Elizabeth Kirby at UC Berkeley maintain that controlled amounts of acute stress can lead to neurogenesis and prime the brain for improved performance. Recent studies on rats found that intermittent stressful events – short-lived, not chronic – triggered the release of fibroblast growth proteins which in turn led to the growth of new neurons. Two weeks later, those mature neurons improved the rats' mental performance. While the study of stress and new cell growth is still being explored, scientists agree that regulated stress creates the optimum conditions for behavioral and cognitive performance. The key is how much stress, how long it lasts, and how well we manage it.

Conversely, chronic stress is ongoing, overwhelming, and often debilitating. People who suffer from chronic stress feel powerless over what they perceive to be unrelenting demands and pressure. Over time, stress becomes the norm and lulls the brain into thinking that this is "just the way life is." When this happens, the brain fails to kick into alert mode. There is no hope for change and no desire to find a solution. Poverty, abusive relationships, or even being unhappy at work can all create the kind of ongoing stress that literally kills people. The vast majority of doctors' visits - more than 90% - are for stress-related illnesses. Chronic stress changes your brain function and structure right down to the cellular level.

With any type of stress, fear, or anxiety, the brain is designed to put your whole body in motion to address it. At the first sign of danger, the brain signals the hypothalamus, pituitary gland and the adrenal glands by your kidneys. Within minutes, the hormones cortisol, adrenaline, and epinephrine are released into the blood stream. With acute stress, we're able to recognize the danger, deal with it appropriately, and stop the flow of chemicals as we return to a sense of calm and peace when the situation is over. Those who suffer from chronic stress are unable to return to a calm and peaceful state, so the brain continues to pump out cortisol, adrenaline, and epinephrine. An overproduction of these hormones makes us more vulnerable to everything from headaches, colds, and disease to impaired cognitive function. The greater the stress, the more cortisol is produced. The more cortisol that is produced, the more damage it does to the body over time.

10 Ways Stress Impacts the Brain and Body

- The overproduction of cortisol can cause high blood pressure and heart attacks. Cortisol constricts the arteries, and epinephrine increases heart rate, which forces the heart to pump harder and faster.

- Cortisol induces the production of glutamate, the neurotransmitter that creates free radicals. Free radicals pierce the brain cell walls and cause them to rupture and die.

- Too much cortisol decreases the production of BDNF (brain-derived neurotrophic factor). This is the body's natural fertilizer designed to keep existing brain cells healthy and stimulate neurogenesis.

- Cortisol shrinks cells and inhibits neural generation in the hippocampus which is responsible for episodic memory, learning, and the ability to regulate emotions.

- Too much cortisol creates architectural changes in the prefrontal cortex; these changes affect executive decision-making and impulsivity control.

- Stress has been called "public enemy #1" because it leads to a host of other health issues including weight gain, heart disease, diabetes, digestive problems, sleeping disorders, skin afflictions, and cancer.

- Stress builds up in your "fear center," or amygdala, and increases the neural connections in this part of the brain which keeps you in a vicious cycle of stress and anxiety. That cycle creates more cortisol, and that cortisol creates more stress and anxiety.

- Stress weakens your immune system making you more vulnerable to anything from a common cold to more severe autoimmune diseases.

- Stress reduces the production of the "feel-good" chemicals, serotonin and dopamine, which causes depression and is linked to addictive behaviors.

- Stress shrinks the brain and kills brain cells.

Read that last line again.
You can be stressed, or you can be smart,
but you won't be both for long.

As an educator, it was important to me that my students were happy in my classroom and genuinely enjoyed learning. What I didn't realize then was that the levels of happiness and stress in the classroom had a direct impact on my students' ability to learn. The same is true for adults. Beyond just making us unhappy, stress physically and chemically changes the structure of the brain and impeding problem-solving, decision-making, creative thinking, productivity and memory.

Think about your daily responsibilities personally and professionally. How often do you make decisions, try to solve problems, try to manage and maximize your time, try not to forget something? By addressing your stress level, you give yourself an edge in all of those tasks. Likewise, if organizational leaders want problem- solvers, innovators, and effective decision-makers on their teams, they'd be wise to assess the stress level in their company culture.

Neurotransmitters – The Short List

To date, neuroscientists have identified over 60 different neurotransmitters and continue to study the roles they play in cognition, behavior, learning, emotions, communication, and even sleep patterns. While they are often classified as having a single or primary function, new research indicates they are multi-faceted, and their interactions can be highly complex. For example, dopamine is frequently called the "pleasure and reward chemical." However, neuroscientists have found that acetylcholine is a neurotransmitter that stimulates the production of dopamine, and the combination of the two chemicals significantly enhance cognition.

The following list explains a few of the more common neurotransmitters that play a significant role in memory, learning, social interaction, mental health, and our overall well-being. We'll explore these in more detail in the coming chapters, but this section will give you a top-line explanation of the primary chemicals and how they affect us.

Dopamine: reward and pleasure

Dopamine is the "I did it!" drug. It helps regulate the reward and pleasure center. Not only does it enable us to enjoy certain rewards, but also motivates us toward the experience. You may get a dopamine rush when you earn a promotion after working hard on a project, lose 10 pounds after sticking to your diet, finish that 5K race after training for months.

People who demonstrate little drive, initiative, enthusiasm or self-confidence may suffer from low levels of dopamine. Studies compared dopamine levels between rats that had two options to get food: an easy option for minimal food and a harder option that produced twice as much food. Those with low dopamine levels chose the easy option for the minimum amount of food. Conversely, the rats with high dopamine levels worked harder to get more food.

People who suffer from addiction may be overproducing dopamine. Because the reward circuit in the brain also includes areas involved with motivation, memory, and pleasure, remembering how good something felt – the high, the rush, the buzz – floods the brain with dopamine and hijacks the prefrontal cortex to go after the source of the pleasure. Over time, consistently high levels of dopamine physically change the brain by desensitizing neurons. This is why some people eventually need more pleasure to get the same "buzz."

Serotonin: mood and well-being

Serotonin (also called the "calming chemical") is primarily responsible for managing mood, anxiety, happiness, and rewarding social behavior. The brain produces it when we feel significant, respected, or important. Recognition from a colleague or family member for doing a good deed or making a major contribution can give you a boost. Exposure to the sunshine, even if only for a few minutes, can also stimulate a release of serotonin.

Low levels of serotonin are associated with depression, although it has yet to be proven whether a decrease in serotonin causes depression or depression causes a decrease in serotonin. A lack of serotonin also impacts appetite regulation, sleep, memory, and decision-making behaviors.

Oxytocin: trust and belonging

Also known as the "hug drug" or "cuddle chemical," oxytocin is the chemical that is produced when we experience the human connection, trust, and sexual arousal. Oxytocin naturally surges during childbirth, breastfeeding, laughter, and orgasm. You can give yourself a boost of oxytocin by giving someone a hug or by simple random acts of kindness.

Low levels of oxytocin have been linked to depression, autism and a range of anxiety disorders. Oxytocin and cortisol are on opposite ends of the hormonal see-saw. As the production of cortisol goes up, it forces the production of oxytocin to go down and vice versa.

Studies have shown that interpersonal touching such as hugging and kissing not only raises oxytocin levels but also decreases cardiovascular stress. Oxytocin is now being tested as a treatment for schizophrenia and autism as well as a potential anti-addictive hormone. Despite the fact that it was discovered in 1952 and listed on the World Health Organization's List of Essential Medicines, we still have much to learn about oxytocin.

Endorphins: pain relief and happiness

Endorphins are commonly known as the chemicals released in response to pain, fear, or stress. Sex, intense exercise, and even some foods like hot peppers (which contain capsaicin) can all generate the feelings of euphoria we get with an endorphin rush. They are typically produced as a response to fear or pain and are found in the regions of the brain responsible for blocking pain. Interestingly, the brain does not have pain receptors, so it doesn't actually feel pain. However, it is the pain processing center for the entire body.

There are over 20 different kinds of endorphins, and some are proven to be as strong as morphine and codeine in our pain management pharmacy. They can also trigger heightened rage and anxiety. If the hypothalamus doesn't receive the endorphins accurately, it will open the cortisol flood gates at the smallest indication of trouble.

Glutamate: facilitates neural communication

Glutamate is the most prominent neurotransmitter in the body (present in over 50% of the nervous tissue), and it's the brain's main excitatory neurotransmitter (increases the electrochemical transmissions of information). Healthy glutamate production contributes to the molecular processes in the hippocampus and cortex. These regions play a large role in learning and memory.

An excess of glutamate can overstimulate the brain and result in *excitotoxicity* which literally means "exciting neurons to death." Excitotoxicity has been linked to strokes, seizures, traumatic brain injuries, and other chronic diseases like Alzheimer's disease and Lou Gehrig's disease.

Gamma-Aminobutyric Acid (GABA): inhibits neural communication

GABA is made from glutamate to naturally counteract its excitatory effects. GABA is an amino acid that acts as a neurotransmitter by inhibiting nerve transmissions when they get overexcited. While glutamate facilitates nerve impulses, GABA works to prevent them from firing. Without GABA, nerve cells fire too often creating the conditions for panic attacks, seizures, and other anxiety disorders. Glutamate is a double expresso and GABA is more like warm milk.

Cortisol, adrenaline, and norepinephrine: stress

Cortisol, adrenaline, and norepinephrine are the three major stress hormones. Adrenaline is the "fight, flight, or freeze" hormone and is produced by the adrenal glands when the brain signals danger. It is responsible for the body's immediate physical reactions to a stressful situation like an increase in heart rate or a surge of energy. Norepinephrine is the arousal hormone produced in the adrenal glands and the brain. This is the hormone that increases your awareness and focus. It also reallocates the blood flow away from areas like the skin, and shifts it toward more essential areas like the heart and muscles. Cortisol is the stress hormone. While adrenaline and norepinephrine are produced in seconds, the release of cortisol can take up to a few minutes. Producing cortisol involves a multi-step process involving two other hormones as well as activity in the hypothalamus, the pituitary gland, and the adrenal glands.

Long-term stress and elevated stress hormone levels are linked to insomnia, chronic fatigue syndrome, thyroid disorders, dementia, depression, and other conditions. People who are unable to produce healthy amounts of cortisol may suffer from adrenal fatigue, low thyroid function, low blood pressure, blood sugar imbalances, and lowered immune function.

Acetylcholine: neuroplasticity, movement, and learning

Acetylcholine was the very first neurotransmitter to be identified. It is primarily responsible for activating voluntary muscle movement by translating intention into action between the neuron and the muscle fibers. Too much acetylcholine is associated with depression, and too little can precipitate dementia. An imbalance of this chemical can have physical effects ranging from convulsions to paralysis. We'll explore the connection between movement and brain function in chapter 6.

Neuroplasticity

Right now, you have about 100 billion neurons transmitting signals in your brain via synapses. The busiest regions of the brain need more oxygen and glucose, so they get more blood. For example, as you read this page, the brain is sending extra energy to various parts of your brain to do that work required to make sense of the letters and words. The occipital lobe is responsible for seeing the printed words and send them to other parts of the brain for processing. The temporal lobe is decoding the words and using phonological awareness to recognize them. The frontal lobe enables you to comprehend the meaning of the words and sentences. The parietal lobe coordinates all of the electrical impulses. It's basically the conductor that links all of the brain regions together so they can work in concert. The more energy those regions of the brain have, the more synapses they can create. And all of this neural activity is continually changing the brain. The more you read, the better you read – not just because you have increased your vocabulary or improved fluency – but because you've exercised all of the brain regions responsible for reading, and over time, that neural activity has changed the structure of your brain and strengthened the parts that need to work together.

Neurons that fire together wire together.

Experiences that are intense, prolonged, or repeated will physically change the anatomy and chemistry of the brain. Neuroscientists refer to this process as experience-dependent neuroplasticity. The brain is a learning muscle. The more we use different regions of that muscle, the stronger those regions get. For example, studies show that taxi drivers in London have developed thicker neural layers in their hippocampus, the part of the brain that helps with visual-spatial memories. The London taxi cab driver is as famous as the black cab he drives. The streets of London are extremely complex and were not constructed with a typical vertical and horizontal grid pattern. To be licensed, London taxi drivers must pass "The Knowledge," a test described as having an atlas of London imprinted in your brain. Because these drivers are required to immediately know the quickest route to your destination without the aid of a map or GPS, they use their hippocampus day in and day out. As a result, that part of their brain physically changes and grows over time.

Similarly, if you make mindfulness and relaxation techniques a part of your daily routine, you'll not only increase the synapses required to calm down in stressful situations, you'll actually change the brain chemistry that enables you to manage stress.

Mental States Create Neural Traits

Emotional experiences such as happiness, worry, love, and stress create synapses that create an imprint in your neural structure. The longer you focus on the experience, the more intense it is, or the more you repeat it, the more likely that mental state will become an imprinted neural trait. If you're in a bad mood (mental state) day in and day out, it will eventually become a characteristic of your personality (neural trait). This is due, in part, to the constant flow of cortisol produced by the negative emotions. This is also perpetuated by the negative feedback loop called the vicious cycle. When the brain is focused on negative events, it gets stuck in a loop that keeps the brain on the lookout for negative things.

We've all had those days that nothing seems to go right. You many not actually be experiencing more bad things on that particular day; you may just be more tuned in to the negatives. The more you focus on the negatives, the more the brain creates neural pathways designed to deal the negatives by producing the neurotransmitters that stop productive brain function to order address the stress. Because your brain is wired to deal with the negative things, you notice them more. On the other hand, the more you focus on the positives, the more your brain will accommodate to enable you to see the positives and produce the good chemicals that create optimum conditions for cognitive function.

The happier you are, the more you'll see the good things. The more good things you see, the more you'll fuel your happiness.

While some people experience more tragedy, loss, or disappointment than others, the real difference between the pessimist and the optimist lies in their brain chemistry. The optimists' brains are wired to focus on the positive experiences because that is what they expect to find, and the pessimists' brains are wired to focus on the negative experiences - because that is what they expect to find.

Reflect and Apply

What did you learn in this chapter that just blows your mind?

Can you identify a time when you experienced an amygdala hijacking?

Can you identify a time when you experienced an endorphin rush?

Do you recognize the stress triggers in your life? Identify two of these triggers and what happens in your brain during these experiences.

1.

2.

Chapter Four
****
The Cognitive Power of Sketch Notes

The World Economic Forum holds an annual meeting in Davos-Klosters, Switzerland where they bring together the world's foremost CEOs, heads of state, ministers, policy-makers, innovators and representatives of civil society. In 2005, the panel consisted of Bill Clinton, Bill Gates, Bono, then-Prime Minister Tony Blair, and then-South African President Thabo Mbeki.

After the panel was over, a journalist from The Daily Mirror wandered across the stage and spotted the notes, scribbles, and doodles of Tony Blair. The reporter, pleased with his discovery, slipped the piece of paper into his folder with the intention of having the hand-written notes analyzed to see what they revealed about the country's leader.

After careful examination by expert graphologists, newspapers and media outlets across the U.K. were eager to announce their findings. Elaine Quigley of the British Institute of Graphologists revealed that Mr. Blair was "struggling to concentrate, and his mind was going everywhere." The Times quoted another expert who maintained Blair was "aggressive, unstable and under enormous pressure." Another announced that he was "not a natural leader."

You can imagine the frenzy that was taking place inside No. 10 Downing Street. After a thorough investigation, one of Mr. Blair's representatives announced that the notes were, in fact, not the scribblings of the Prime Minister. Rather, the notes belonged to Bill Gates who was seated next to Blair at the forum.

Well, that's a relief! The aggressive, unstable, unfocused doodler wasn't the Prime Minister. Instead, those words described the world's richest and greatest philanthropist.

A year later, David Greenberg, a history professor at Rutgers University, published a book called *Presidential Doodles: Two Centuries of Scribbles, Squiggles, Scratches & Scrawls from the Oval Office*. As he would share in the book, many of our leaders throughout history have been avid doodlers. Theodore Roosevelt doodled animals; JFK doodled sailboats, and Ronald Reagan doodled cowboys.

Albert Einstein, Steve Jobs, John Lennon, and Leonardo da Vinci all doodled and sketched brilliant insights and discoveries. Walt Disney started doodling and sketching at the age of 4 and was well known for his daydreaming. Despite all of these masterminds who sketched, scribbled, and visualized their thoughts, many people still associate doodling with a lack of focus rather than with the creative process.

Several years ago, after reading Sunni Brown's *The Doodle Revolution* and practicing with *The Sketchnote Handbook* by Mike Rohde, I decided to test the science out for myself. I sketched grocery lists, blog ideas, and random thoughts. My sketch notes gradually improved, and I was finding value in this new visual format. I began to apply it to my daily work. I remember sitting in a meeting at the "grown-up table" sketching my notes. A colleague leaned over to me and sarcastically asked, "Are we *boring* you?"

He was a bit higher on the food chain than I was, so I chose not to "school" him on the cognitive power of sketch notes. I smiled politely and said nothing, but on the inside, I was screaming, "Hey buddy, my sketches and doodles are actually preventing you from boring me!" (Perhaps I should send him this book with a doodled inscription.)

Most people assume that when we're bored, the brain is inactive. The opposite is actually true. The brain is designed to actively process data; neurons were built to fire. When the brain lacks stimulation, its default is to go out looking for something better to process. Hence, the daydream. The power of the doodle is that it provides just enough stimulation to keep the brain engaged with the information at hand so that it doesn't drift off in search of something better. Recent studies now show that doodling can actually improve focus, grasp new ideas, and stimulate deeper insights. Put another way, doodling engages the whole brain. Some researchers theorize that doodling helps the brain maintain a baseline of activity in the cerebral cortex when outside stimuli are absent. In a 2009 study published in Applied Cognitive Psychology, people who were encouraged to doodle while listening to a list of names were able to remember 29% more of them later when surprised with a quiz.

Gabriela Goldschmidt, a researcher on learning techniques of design, summarized this concept effectively when she said, "a doodle can spark dialogue between the thinking mind and the drawing hand and the seeing eyes." Drawing information rather than writing it visually engages multiple neurological networks simultaneously. In other words, the whole brain actively engages with what it "sees." Doodles and sketch notes enable you to create rich visual maps of what you see, hear and think. In doing so, you're integrating different regions of the brain and tapping into deeper cognition.

Sketching, doodling and drawing information enables the brain to switch into free-flow mode and break away from constrictive habits of thinking. Brainstorming, innovation, and troubleshooting all require creative thinking processes. When we can break free from habitual patterns of thought, new ideas will emerge.

The enemy of creativity is not a lack of imagination;
it is a commitment to the prior art.

Mind the Gap

Doodles and drawings are not just valuable to the individual; they can be an effective way to improve communication and collaboration within departments and across the organization. Perhaps the biggest organizational benefit from sketch notes is illustrated by the well-known tree swing cartoon. Anyone even remotely engaged in software project management has likely seen and shared this illustration. But its origins are a bit of a mystery. Some claim that the cartoon originated in the 1960s in the United Kingdom. However, one of the earliest documented versions that I could find was printed in the University of London Computer Centre Newsletter No. 53, March 1973 with the caption, "Acknowledgements to Unknown Author." While we may not know exactly who is responsible, people have adapted the images to their own discipline and industry for more than 40 years.

The concept is an obvious illustration of the pervasive communication problems facing so many organizations today. Workplace studies show that the average employee spends more than 75% of his/her time on team-based tasks. Those organizations that improve their communication processes and illuminate gaps in understanding will save valuable resources. The key is that those gaps and misconceptions have less to do with the project and more to do with hidden assumptions. Whether your team is tasked with building a tree swing, developing a computer program, or creating a marketing initiative for a new product, we all come to the table with hidden assumptions. Those assumptions can be both inaccurate and costly.

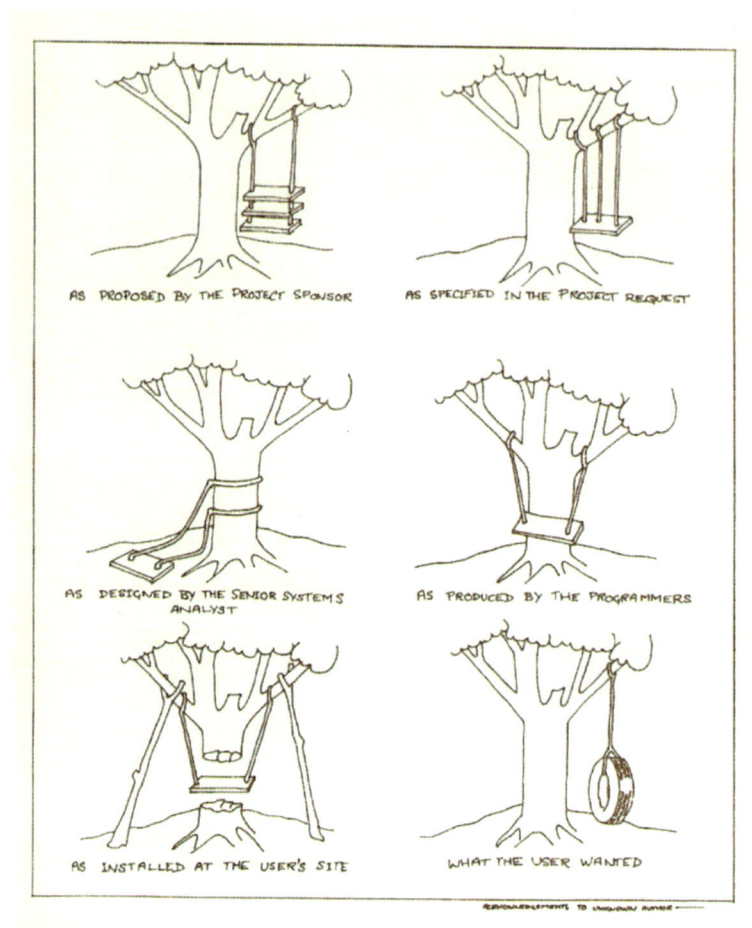

[Tree swing cartoon with six panels:
- AS PROPOSED BY THE PROJECT SPONSOR
- AS SPECIFIED IN THE PROJECT REQUEST
- AS DESIGNED BY THE SENIOR SYSTEMS ANALYST
- AS PRODUCED BY THE PROGRAMMERS
- AS INSTALLED AT THE USER'S SITE
- WHAT THE USER WANTED]

From the *University of London Computer Centre Newsletter* No. 53, March 1973

While many people recognize the tree swing cartoon and the implications of hidden assumptions, far fewer understand the value of sketch notes as a way to overcome them. To demonstrate the concept, I ask participants in my leadership workshops to engage in a very simple activity. It begins with a few questions:

How many of you know how to make toast?
Every hand goes up.
How many of you are confident in your abilities to teach someone else how to make toast?
Every hand goes up (and a few eyes roll).
How many of you feel confident that everyone on your team knows how to make toast?
Again, every hand goes up (and inevitably someone will mutter, "If they can't make toast, we're in trouble!").
If you tasked them with describing the steps to make toast, how long should it take them?
The responses typically range between 2 minutes and 5 minutes.
So, If I gave you 15 minutes to sketch out the steps for making toast, that would be enough time to create clear and accurate instructions?

I ask these leaders to work in small groups to sketch out the steps for making toast. I tell them to approach this task as if it were an instructional guide that others in the organization would follow. Consider three different examples of this simple task created by employees within the same organization.

This group made the assumption that the bread and toaster are readily available. In five simple steps, the team outlined how to "make the perfect toast" with optional toppings.

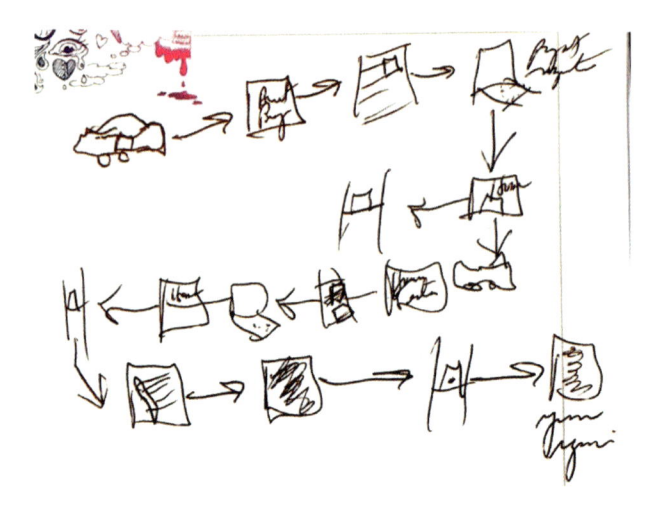

This group started by getting in the car, going to a store (Best Buy) to purchase a toaster, then going to the store to purchase the bread. After returning home, one would have to plug in the toaster, put the bread in, wait for it to heat up and toast the bread. Finally, remove the toast, add butter (no other optional toppings) and enjoy ("yum yum").

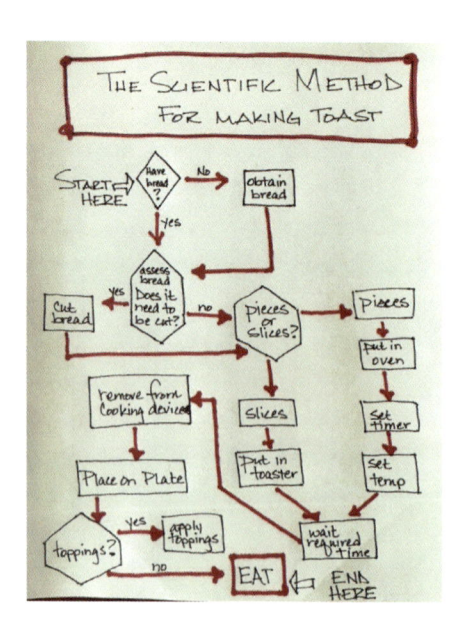

Another group turned the simple task of making toast into a "scientific flow chart" by including a complex series of questions and options based upon the answers to those questions. It would appear that they had assumed nothing, however in the discussion that followed, they said if they had more time they could have been more thorough adding options for various cooking times, temperature and toppings.

This activity is always incredibly insightful. Some groups begin the instructions by growing the wheat to make the bread, while others begin with sliced bread, and still others make the toast in a pan on the stove. But it isn't until they start sketching out the steps that they recognize the gaps.

"Wait, where did you get the bread?"
"The store."
"But where did the store get it?"
"What if you don't have a toaster?"
"I make my bread in the oven."
"We need to provide an option for gluten-free bread."
"I slice my toast diagonally before I eat it."

The point is we all have hidden assumptions. Sketching out the process not only engages the whole brain for deeper cognition, it also helps people – individually and collectively – see how the details fit together in the big picture, identify the hidden assumptions, and address the gaps.

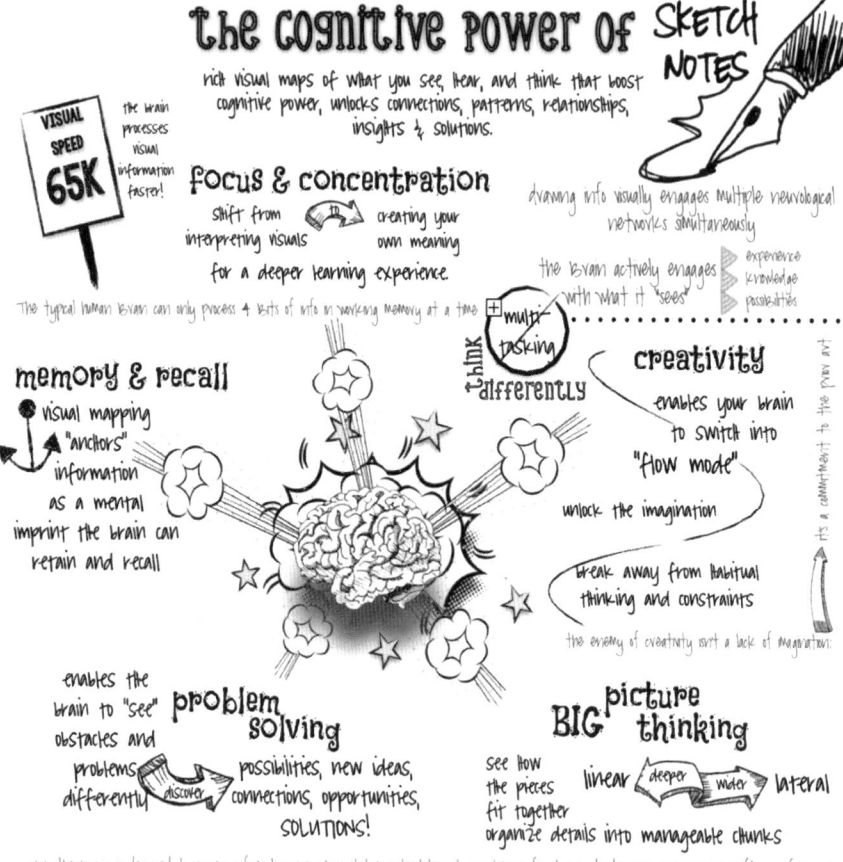

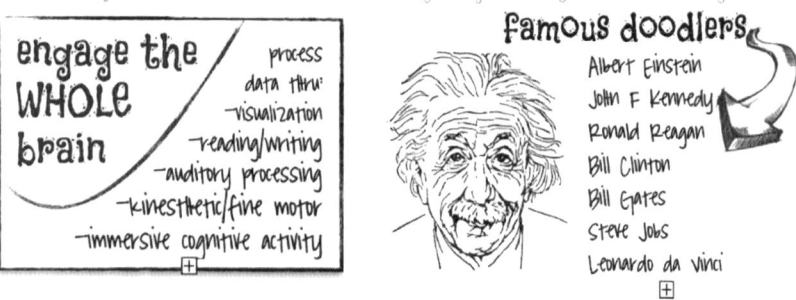

The Cognitive Power of Sketch Notes

rich visual maps of what you see, hear, and think that boost cognitive power, unlocks connections, patterns, relationships, insights & solutions.

VISUAL SPEED 65K — the brain processes visual information faster!

Focus & Concentration
shift from interpreting visuals → creating your own meaning for a deeper learning experience.

drawing info visually engages multiple neurological networks simultaneously

the brain actively engages with what it "sees" → experience, knowledge, possibilities

The typical human brain can only process 4 bits of info in working memory at a time

multi-tasking → think differently

Memory & Recall
visual mapping "anchors" information as a mental imprint the brain can retain and recall

Creativity
enables your brain to switch into "flow mode"

unlock the imagination

break away from habitual thinking and constraints

the enemy of creativity isn't a lack of imagination

it's a commitment to the process of art

Problem Solving
enables the brain to "see" obstacles and problems differently → discover possibilities, new ideas, connections, opportunities, SOLUTIONS!

BIG Picture Thinking
see how the pieces fit together

linear → deeper → wider → lateral

organize details into manageable chunks

*** the brain is designed to process information *** when it lacks stimulation, it goes looking for it *** daydreamers beware *** shift your focus ***

Engage the WHOLE Brain
process data thru:
- visualization
- reading/writing
- auditory processing
- kinesthetic/fine motor
- immersive cognitive activity

Famous Doodlers
- Albert Einstein
- John F Kennedy
- Ronald Reagan
- Bill Clinton
- Bill Gates
- Steve Jobs
- Leonardo da Vinci

Reflect and Apply

"A doodle can spark dialogue between the thinking mind and the drawing hand and the seeing eyes."

Use this space to doodle a quick summary of something you've just learned. It could be something from this chapter or something new you've learned from a friend or colleague. Use boxes, shapes, arrows, and other figures to create a visual map of the information.

"Live life to the fullest. You have to color outside the lines once in a while if you want to make your life a masterpiece. Laugh some every day."

Chapter Five
****
Laughter is the Best (Brain) Medicine

In July of 1955, a brand new theme park opened in Anaheim, California. An ambitious ten-year-old boy ventured in with a mission. He left that day as the newest employee at Disneyland. He returned the next day to sell guidebooks, but after a few months he graduated to working in the magic shop where he discovered he had a natural talent for public performance. He carefully studied the magicians, learned magic tricks and eventually earned his own time slot to perform. While he loved to fascinate people with magic, he discovered his real joy was making people laugh.

That summer, his quest to become a professional comedian began. He wrote jokes and continued to work on his delivery with friends and family. The venue grew from the sofa in his living room to the talent show in his high school gymnasium. By the time he graduated from high school, he had developed enough material to perform in local comedy clubs and bars and joined a comedy troupe at Knott's Berry Farm. At age 23, he won an Emmy Award for his writing work on *The Smothers Brothers Comedy Hour*.

In 1975, he was invited to perform on *Saturday Night Live*. SNL's audience went wild. He became a frequent host of the show, and each time he appeared the television audience jumped by a million viewers. He brought the "air quotes" gesture to life, and established the national catch phrase "Excuuuuuse me!"

He was just a wild and crazy guy.

His name was Steve Martin, and he's made millions of people laugh. He's also made millions of people a little smarter.

Did you hear the one about...

We know that it feels good to laugh, and we associate laughter with "fun times" and positive energy. But, from the scientific standpoint, laughing and smiling has been clinically proven to have a positive effect on physical, emotional, and social health and well-being.

We all know how jokes work. They defy our expectations. We find humor in the incongruity between the setup and the punchline.

A young girl says to a gymnastics instructor, "I want to be an Olympic gold medalist like Simone Biles. Can you teach me?"

The instructor replies, "I don't know... how flexible are you?"

"I'm available after school on Mondays and Fridays."

You get it. But, the brain begins to change as it anticipates humor – even before the punchline.

A priest, a politician, and a clown all walk into a bar.

The incongruity of the mental image you conjure up when you hear that immediately puts your brain into motion. You envision the setup, you know it's a joke, and your brain prepares for it.

The bartender says, "What is this, some kind of joke?"

Maybe you laugh or maybe you groan. Either way, your brain is engaged in neural activity even as it prepares to process the joke. Even the simplest joke like the one above engages several brain regions. First, the frontal lobe processes the incongruent information and determines whether it's worthy of an emotional response. It's passed along to the supplementary motor area located in the cerebral cortex which controls the movements associated with laughter. Finally, the limbic system springs into action to mix up a healthy dose of oxytocin, serotonin, and dopamine compliments of the pleasure center.

Even just anticipating laughter can reduce stress

hormones such as cortisol and epinephrine and boost the kinds of hormones that protect the immune system. According to a study conducted at Loma Linda University under the direction of Dr. Lee Berk, the interaction between the brain, emotional behavior, and the immune system suggest that the mere act of seeking out opportunities to laugh will impact our physical health.

Is there a difference between a belly laugh and a chuckle? It turns out there is a big difference. Studies show that engaging the diaphragm with ANY type of deep breathing – especially hearty belly laughing – immediately engages the parasympathetic nervous system and sets off a chain of events throughout the body. The parasympathetic nervous system directs all of the other body systems to slow down so the stress hormones can take a break and the "feel good" hormones (endorphins) can take over. Once all of that happens, blood pressure drops, heart rate slows and the entire body gains the benefits as happiness replaces anxiety and stress. It's the perfect domino effect.

While scientists have studied brain activity in the emotional states of fear, anger, and depression for many years, the exploration of neural activity stimulated by laughter has exploded over the last decade or so. Neuroscientists have discovered valuable insights between depression and laughter. For example, people who suffer from depression have an imbalance between the frontal lobe and the pleasure center of the limbic system. Both of those areas are essential to the neural activity associated with laughter. Is it possible that in a state of depression, we laugh less because we are less able to laugh? Neuroscientists are now exploring laughter as more than just a response to something funny.

Laugh and the Whole World Laughs with You

Have you ever noticed that television sitcoms use laugh tracks? The reason for this is – you guessed it – neuroscience. The scientific term is *emotional contagion*. We've known for some time that people tend to mimic behavior, gestures, body language in conversations. Studies have shown that people can "catch" fear or anxiety when interacting with others exhibiting those emotions. Likewise, seeing someone else laugh or smile triggers a corresponding emotional response. In addition, just hearing different emotional sounds such as laughter, cries, or screams can trigger a reaction in the premotor region of the brain that prepares the listener's facial muscles to respond accordingly. Neuroscientists attribute this to mirror neurons. Mirror neurons are a subset of brain cells that fire when we perform an action as well as when we see others performing a particular action. More recent research has explored mirror neurons as the psychological mechanism that enables us to understand the thoughts, actions, and intentions of others by allowing us to "feel" another's emotions rather than just observe them.

Shawn Achor, the author of *The Happiness Advantage*, describes how a group of hospitals in Louisiana used this as a social support construct to reinforce the notion that we're all connected, and the emotions and attitudes we project are contagious. More than 11,000 doctors, nurses and other staff members were trained on the power of the smile. They implemented a particular behavioral strategy called the "*10/5 Way*" to establish positively contagious interaction. Employees were asked to make eye contact and smile at people who were within 10 feet of them. When they were within 5 feet of others, the expectation was to make eye contact, smile, in addition to a verbal greeting.

Six months after they initiated the program, they observed happier employees, more satisfied patients and an increase in referrals, all because of a simple social interaction. Imagine that…a free, one-second behavior requiring no special training or fancy equipment made a happier workplace!

To date, there are no widely accepted studies that confirm or explain how mirror neuron activity supports cognition. What we do know is that smiling and laughter are empathetic, validating social experiences that bind us. In fact, we are 30 times more likely to laugh when we're with other people. Laughter is an emotional response that makes us feel closer, more trusting, more connected. Another round of oxytocin, endorphins, dopamine, and serotonin, please!

Mental benefits aside, laughing and smiling can actually make you live longer. Remember all of those negative ways that stress physically impacts the body? Laughter and smiling are on the opposite end of the spectrum, and they have been proven to counteract stress. Both laughter and smiling relax the body by reducing heart rate and blood pressure. When the body can relax, the white blood cell count increases enabling the immune system to operate more efficiently. Cortisol levels decrease and let the good chemicals balance your mood into a state of well-being. Laughter also induces the release of endorphins which increases your pain threshold and reduces the pain you feel.

Workplace studies have also shown that the extent to which we smile and laugh at work impacts how people perceive us professionally. People who smile more frequently appear more confident, trustworthy, and collaborative. Smilers are more likely to get promoted and higher salary increases than their nonsmiling peers (all other things being equal). Because laughter changes our state of mind and promotes positivity, it can also enable us to see challenges or problems from a different perspective and enable us to find creative solutions.

If all of that isn't enough to turn that frown upside down, smiling may shave years off your face. The face muscles you use to express anger, stress and anxiety pull your face downward, while the muscles you use to smile lift your face making you appear thinner and younger. In one study using over 2000 photographs of faces with various expressions (angry, fearful, disgusted, happy, sad, or neutral), participants were asked to guess the age of the person in each photo. The estimated ages of the neutral faces were the most accurate while participants underestimated the ages of the happy or smiling faces by an average of 3 years. Furthermore, we tend to judge people with sad faces as being heavier.

Forty years ago, it would be hard to find a scientist who prescribe laughter and smiling as a strategy for overall well-being. Today, researchers are beginning to understand just how emotional states impact physical health.

So, take the time today – TODAY – to tell a joke, make someone smile, or look for a reason to laugh.

Reflect and Apply

What was the biggest "AHA!" you learned in this chapter?

List two specific recent situations where humor or laughter could have made people feel more connected or improved the outcome.

1.

2.

Identify one way that you can make laughter work for you *tomorrow*. Be specific: who, where, when, how and why?

"Do not grow old, no matter how long you live. Never cease to stand like curious children before the Great Mystery into which we were born."

Chapter Six
****
Get Your Groove On

Steve Jobs made the "walking meeting" a normalcy at Apple. According to his biography, the length of the walk determined how serious the conversation was.

Harry Truman's daily routine consisted of waking up at 5:00 AM, getting dressed in a suit and tie and taking a vigorous 1-2 mile walk before beginning his work day.

Mark Zuckerberg prefers his first meetings with people to be walking conversations because he finds it less distracting than office meetings.

Charles Darwin installed a gravel path similar to a race track that he would walk to think through problems. The number of laps he walked depended upon the difficulty of the problem. (That guy would have been the Fitbit poster child of the 1800s!)

Fitness gurus maintain that exercise is the key to happiness. You're healthier, you're less stressed, you rock that great outfit, and you're one workout closer to that swimsuit. All of that feels great! Serious runners have experienced a rush of endorphins, also known as the "runner's high." But few people include physical activity into their daily routine for the most compelling reason - brain health. Movement benefits the brain long before it impacts the body. The connection between mind and body has been explored in depth for over 100 years. Now, advances in technology enable us to peek inside the brain to see what actually happens when we move the body. As the title of this chapter already told you, gross motor movements spark brain activity. Using fMRI scans, we can see a significant increase in brain activity after just a 20-minute walk. The bonus is that this particular type of activity is also associated with happiness. Smarter and happier? Yes, please!

The brain demands more energy than any other organ in the body. The more active it is the more fuel it needs. However, despite the fact that it's a fuel hog, it neither produces nor stores the oxygen and glucose it needs for neural activity. The brain relies on the body to keep the tank full. When it has the appropriate amount of fuel, it changes at a molecular level through neurogenesis, the release of "feel good" neurotransmitters, and the production of BDNF, a neurotropic factor that is a little like Miracle-Gro for the brain.

While there is no such thing as one single movement center in the brain, the cerebellum controls most of our motor functions. Surprisingly, the cerebellum takes up just one-tenth of the brain by volume, yet contains more than half of the brain's neurons. While information travels to and from the cerebellum, most of the neural circuits are travel outward transmitting impulses to the visual system, sensory cortex and other parts of the brain. The cerebellum also works closely with the vestibular (inner ear) to help us coordinate movement and turn thoughts into actions. All of this activity monitors and regulates incoming sensory data and increases our ability to attend and focus. It also creates the optimal environment for neural plasticity.

Educational experts are finding great academic gains when incorporating movement in classroom learning experiences. Even though education has been traditionally a "sit and learn" environment, the "movement" about movement began over 100 years ago with Dr. Maria Montessori.

> *"One of the greatest mistakes of our day is to think of movement by itself as something apart from the higher functions. Mental development must be connected with movement and be dependent on it."*
> *-Dr. Maria Montessori*

Regular physical activity is essential for maintaining brain health for people of all ages. One of the first scientifically controlled studies showing the power of exercise examined how a brisk walk 30-40 minutes per day three times per week might impact cognitive decline in senior citizens. The result was the equivalent of stopping the aging brain clock by up to two years. Additional studies have been conducted to identify how different types of exercises impact brain function. In a study conducted at the University of British Columbia, researchers found that regular aerobic exercise that activated the heart and the sweat glands increase the size of the hippocampus, the region involved with memory and learning. Exercises such as resistance training and muscle toning did not have the same impact on the hippocampus. Scientists expanded on that work with additional studies. They found that the kinds of physical activities that integrate different parts of the brain such as coordination, balance, and rhythm are the best brain workouts. Cycling, for example, enhances brain function more before, during, and after a workout than running for the same amount of time. Ballroom dancing, another activity with both physical and mental demands, has a higher impact on cognitive functioning over exercise or mental tasks in isolation.

Any physical activity that gets the heart pumping results in increased blood flow to the brain. In a 2016 study published in Frontiers in Aging Neuroscience, researchers sought to determine the effects of the increase in blood flow had on the hippocampus and memory function. The subjects of the study were competitive runners between the ages of 50 and 80 who all ran at least 35 miles per week. The athletes were asked to engage in as little physical activity as possible for ten days. The results showed a striking decrease in blood flow to most regions of the brain and significantly less to the hippocampus.

Physical movement fuels brain cells, stimulates cell growth and promotes the dendritic branching process. And let's be honest, who couldn't use a few more brain cells? The more brain cells we have, the more branches we can grow. Even just a quick 20-minute cardio workout before taking on a mental task will fire up the brain cells and get the creative juices flowing. But, you don't have to work up a sweat to get your brain fired up. While aerobic exercise that gets your heart pumping is ideal, it's not always practical to do in the middle of your workday – often times when you need it the most. Here are a few keys to understanding the impact of physical activity on neural activity and a few simple ways to build them into your day.

Pump the heart; feed the brain

The fuel station is in the heart. The faster the heart beats the more energy it pumps to the brain. The more energy the brain has, the better it functions. But typically, when the brain is engaged in a cognitive task and needs fuel the most, the body is idle at a desk or in front of a computer. Simple movements such as lunging, walking, or simply stretching up as high as you can and then squatting down the to the ground a few times can raise your heart rate by up to 10% and send a little extra oxygen and glucose to refuel the brain.

A brisk walk in the morning is a great way to jump start your brain. Not only will it give you an edge with mental stresses throughout the day, a boost of fuel to the brain early in the day will also sharpen your ability to retain new information and process more complex tasks.

Create a change of scenery

The technical term is "episodic encoding," but the basic premise is that when we learn new information, the prefrontal cortex generates a map (people, places, emotions, the context of the information, etc.) documenting the meaningful data before sending it off to the hippocampus. The more detailed the map is, the easier it is to understand, apply and recall the information later.

Have you ever noticed that people tend to sit in the same seats at routine meetings? We are creatures of habit, and we look for that which feels familiar. It's learned behavior that started back in grade school. But it's a behavior that could be stifling to creativity and problem-solving. The next time you walk into the conference room, make it a point to sit in a different seat than you did during the last meeting to add more data points to your brain map.

If you find yourself stuck on a problem, try working on it somewhere else. Grab a notepad, find a park bench and think it through there. Or head over to the nearest Starbucks. You don't even have to leave the room; just changing your position in the room can add details to your brain map and impact the way you process information to give you greater insight.

Give it a break

Our brains continually take in a lot of information. The hippocampus evaluates that data and determines whether we save it or delete it. The time the brain allocates to sort and organize that information is critical. When we overload the hippocampus, it's harder to evaluate data or receive new information. Short intervals that break up the constant flow of incoming data can be enough to let the brain catch up and continue to process information efficiently. Step away, grab a bottle of water, do a lap around the building, and hit the "mental pause button."

Stand and deliver

We don't always have the freedom to move about or take a brief walk at work. Recent studies have shown that just the mere act of standing for short periods of time can give the brain a boost. Standing desks have gained traction in both academic as well as professional settings, largely due to studies that have shown notable improvements in cognitive function.

Even if you don't have a standing desk, take a break from sitting every so often. Set an alert on your computer or your phone as a reminder. Grab a notepad and stand up to think or jot down ideas. While it isn't a replacement for exercise, it is one way to increase blood flow to the brain and boost focus, memory, and cognition.

Move to motivate

The brain produces dopamine to mediate pleasure and rewards. These chemical energizers also stimulate motivation, attention, and cognition. Certain repetitive gross motor movements stimulate the production of dopamine. Feeling sluggish, foggy, or distracted? Or maybe you tend to hit "the wall" at about 2:30 or 3:00 in the afternoon. On those days, pay attention to how long you've been stationary or parked in front of the computer. Climb a set of steps, do a few lunges, or take a 5-minute brisk walk to poke your limbic system and fuel the brain.

Put it to the test

From the classroom to the conference room, learners of all ages can improve cognition, memory, and creativity through physical activity. The brain was just not designed to work with an inactive body. Experts maintain that it takes about 150 minutes of moderate physical activity per week to impact brain health. If 150 minutes seems like a big number, break it down. Set a realistic goal for to be physically active every single day with a reasonable number of minutes each day conducive with your schedule. Maybe something like this:

Day	Activity/Minutes	Time
Monday	brisk walk/15 min.	during lunch
Tuesday	brisk walk/20 min.	before work
Wednesday	bicycle/15 min.	before work
Thursday	brisk walk/25 min.	after work
Friday	brisk walk/15 min.	during lunch
Saturday	brisk walk/30 min.	in the morning
Sunday	bicycle/30 min.	in the afternoon

All of those short chunks of time add up to 150 minutes! The key to remember is that a few minutes is better than nothing. Start with 10 minutes a day and gradually increase the number each week. Ask a co-worker to walk with you. If you can't get your walk in at lunch, make it up in the evening. Mix it up with other activities you enjoy like swimming, tennis, or dancing. Just make movement a part of every day. Your body and your brain will thank you.

Reflect and Apply

List two key nuggets you learned from this chapter.

1.

2.

Identify two realistic ways in which you can incorporate more physical activity into your daily routine.

"Do not grow old, no matter how long you live. Never cease to stand like curious children before the Great Mystery into which we were born."

Chapter Seven
****
Surprisology

My sister lives in Ohio. Each time she visits, she hides little sticky notes all over my house. I never know how many notes I'll find or where I'll find them. I fight the urge to go looking for them. Rather, I've learned how to find joy in *discovering* them...inside the egg carton in the refrigerator, in the laundry room cabinet, on a random page of my planner. One day, while digging through a drawer for my green socks, I'll find a note that says, "I hope you're having a great day!" Another day, I'll pull out the muffin tin to find a note that says, "I wish I were there to share a sweet treat with you!" Every single time I find one of those notes, I smile or laugh. Every time that happens, my sister gives me a little boost of dopamine from 1,200 miles away!

"Surprise!" That single word can create excitement for some people and anxiety for others. A friend of mine loves the element of surprise – anything from a little treat tucked into his briefcase or a spontaneous trip to Paris. He embraces the unknown, but he's in the minority. By design, surprises catch us off guard and unprepared. Most people will admit that they prefer predictability to surprise – even if the surprise is a good one.

Neuroscientists have discovered that surprise is one of the most powerful and least understood human emotions. We are wired to notice novelty. In fact, despite what you may think, the brain LOVES novelty. This is why we laugh at jokes, enjoy movies or books with an unexpected twist at the end, visit a new restaurant, or go somewhere we've never been for a vacation.

The brain's pleasure center (or the nucleus accumbens) lights up like a Christmas tree when you experience something good that you didn't expect. Not only do you get a nice boost of dopamine, but the brain also releases noradrenaline – the neurotransmitter responsible for focus and concentration. (Yes, this is another reset button for the brain.) When we are surprised by something – either good or bad – our brain actually stops much of the cognitive activity to focus on the surprise and find meaning in it. Think of surprise as what makes an ordinary event a memorable one – either from delight or disappointment. But, here is the key: good surprises increase the dopamine levels and bad surprises decrease the dopamine levels. Depending on the intensity of "bad surprise," the body may even begin produce cortisol, adrenaline, and norepinephrine for a "fight, flight, or freeze" response.

Authors Tania Luna and Leeann Renninger explore the science behind surprise in their work as "surprisologists" at Surprise Industries and LifeLabs New York. Their book, *Surprise: Embrace the Unpredictable and Engineer the Unexpected* is something of a how-to book about making surprise work positively both at work and in your personal relationships. Luna and Renniger maintain that novelty or surprise intensifies our emotions by about 400 percent. This concept of *emotional intensification* explains why receiving flowers on your birthday is nice, but getting "just because" flowers on some random Tuesday feels so much different. The gift is the same, but the surprise flowers elicit a much more intense response than the birthday flowers.

Conversely, bad news feels much worse when we are surprised by it. Consider two people who lose their jobs. One person is let go when a number of positions are eliminated as a result of an acquisition. The second person is fired without warning. The outcome is the same for both people; they've both lost their jobs. In the case of the acquisition, people typically anticipate personnel changes. On the other hand, being blindsided with absolutely no warning is emotionally much more devastating.

When we begin to not only understand but embrace the element of surprise, we're better able to thrive when we are uncertain or don't have complete control of things. Moreover, the element of surprise is one of the secrets to personal growth and human connection. Surprise enables us to make an ordinary experience more meaningful. The perfectly planned experiences may give us satisfaction, but they aren't the most memorable. The surprises in life are imprinted in our memory banks.

Perhaps the best part of surprises is that they don't have to be huge to make a huge impact. The unexpected thing – a random act of kindness, a handwritten thank you note from a colleague or the little notes that my sister hides in my sock drawer – can create positive chemical changes that reset your brain, create new neural pathways that make you a little smarter and a little happier. Another side effect of delightful surprises is that all that happiness is contagious. You already know about emotional contagion from the chapter about laughter. The same concept applies with surprise. Because we're social creatures, we tend to "catch" the emotional states of those around us. Interacting with people who are stressed and anxious significantly impacts your own mental state and releases stress hormones in *your brain*. Likewise, you experience the surprise of others vicariously and you feel the same connectedness, sense of belonging, and emotional. Your brain releases the good stuff like oxytocin, serotonin, and dopamine.

Happiness is contagious; so is stress.
Which would you rather catch?
Would you rather be responsible for spreading happiness or stress?

Whether you love surprises or hate them, it's not that complicated to make them work for you and others around you. When you understand how surprise works in the brain, it's even sweeter. Here are a few simple ways to both inspire and embrace surprise.

Practice of the art of vulnerability.

Brené Brown said, "Vulnerability sounds like truth and feels like courage. Truth and courage aren't always comfortable but they're never weakness." As much as I love this quote, vulnerability can be scary. When we reveal our authentic selves, we're open to judgment, rejection, and criticism. But we also enable freedom and opportunity when we learn to let go. Research has shown that demonstrating vulnerability is humanizing, endearing and incredibly powerful in helping us connect with others. Those connections give us purpose and meaning. Nurturing your sense of vulnerability and learning how to laugh at yourself is one way to allow life to surprise you. It is also a powerful way to strengthen your most important relationships.

Nurture curiosity.

The degree to which we are curious has a direct correlation with personal growth and our ability to connect with others. Curiosity opens your brain to the notion that you're okay with not knowing and allows you to explore concepts to understand rather than to judge. It's the cornerstone of learning and discovery, and it requires you to let go of control and embrace unpredictability.

Be unpredictable.

We tend to be creatures of habit because it's "comfortable." Most of us gravitate toward familiar faces, spaces, and places. Stepping into the unknown, even for the most mundane tasks, may require intention and focus. But, a little unpredictability can often lead to new insights, new perspectives, and some amazing "a-ha" moments. Break away from your routine, take a new route to work, invite someone outside of your social circle to lunch, or plan an impromptu one-tank-trip to a nearby city you've never explored before. Be on the lookout for the unexpected.

Surprise and delight people around you.

Look for little ways to make someone else smile. A kind unexpected act or gesture doesn't have to cost a lot of money or time, and can give the giver and the receiver a nice boost of dopamine. Think of simple ways to surprise and delight others like leaving a "thank you" note and a tea bag on a coworker's desk. Slip a note into your spouse's jacket pocket or the console of his or her car. Write a knock-knock joke on the napkin you put in your child's lunchbox. Bake some homemade muffins for the elderly lady next door.

Take a minute today to figure out a way to surprise someone in your corner of the world. Don't stop there. *Make it an intentional part of your thought process.*
A quick note, a reason to laugh, or an invitation to lunch… how delicious to know that small surprise will light up someone's pleasure center!

Reflect and Apply

Describe a time when you experienced positive "emotional intensification."

Describe a time when you experienced negative "emotional intensification."

Think about the people in your corner of the world who you could treat with a surprise. How can you surprise them?

1.

2.

How can you incorporate "surprisology" into your daily routine (either for yourself or others)?

"It is every man's obligation to put back into the world at least the equivalent of what he takes out of it."

Chapter Eight
****
The Grateful Brain

"You may never have proof of your importance, but you are more important than you think. There are always those who couldn't do without you. The rub is that you don't always know who."

That little nugget is by Robert Fulghum in his book, *All I Really Need to Know I Learned in Kindergarten*. By the time we can tie our shoes, we know the importance of saying "thank you." It's one of the first social courtesies we're taught. Somewhere between the kindergarten classroom and the rat race of life, the practice of gratitude often gets lost.

If good manners aren't enough, recent studies have proven that sincere expressions of gratitude can have a significant impact on healthy brain activity as well as physical and psychosocial health. Researchers at the National Institutes of Health examined the neural activity and blood flow in various regions of the brain when people experienced gratitude. They found that greater levels of gratitude generated increased activity in the hypothalamus. Remember, the hypothalamus is responsible for some pretty important body functions such as eating, drinking, sleeping, metabolic activity and managing stress levels. In addition, feelings of gratitude directly activate the limbic system and trigger a release of dopamine. Dopamine is the reward chemical, but it is also responsible for initiating the action to get that good feeling again. It's your brain saying, "Oh… that felt good! Do that again!"

That shot of dopamine is also what sends your brain into what neuroscientists call the *virtuous cycle*. As complex as the human brain is, it has a one-track mind. It likes to focus on either positive stimuli or negative stimuli but not both at the same time. When the brain is focused on positive events, the natural tendency is to stay in that positive loop until a negative experience ultimately intervenes and breaks the cycle. Conversely, the brain can also get stuck in a negative loop called a vicious cycle. This is what I like to call WMS or "Why Me Syndrome." When the brain gets trapped in the vicious cycle, it tends to focus on the negatives. "The traffic made me late for work, someone took my parking place, I spilled my coffee, it's raining and I forgot my umbrella, my boss is a jerk…" There may be many positive things going on, but the brain is too busy processing all of the negatives to notice them.

The brain also has a natural tendency to look for things that prove what it believes to be true. It's called *confirmation bias*, and it can be both friend and foe. For example, if you get up in the morning and believe that you're going to have a miserable day, your brain will search for evidence to prove you right. Likewise, if you start your day with the belief that life is good, your brain will search for evidence to confirm that worldview. The outlook you choose determines whether you'll get stuck in the virtuous cycle or the vicious cycle. The only way to get out of the vicious cycle is to intentionally point your brain in the other direction.

In gratitude study conducted by researcher and author, Robert Emmons, participants were randomly assigned to one of three groups. All three groups were given a weekly journaling assignment for a total of ten weeks. One group was asked to record five things that happened during each week for which they were grateful. The second group was asked to record five obstacles or challenges they experienced each week. The third group was instructed to record five events from the week that had an impact on them but were not told whether to focus on positive or negative events. The participants in the first group who recorded positive expressions of gratitude reported fewer physical aches and pains, felt better about their lives in general, and were more optimistic about the upcoming week compared to the other two groups. They also exercised an average of 1.5 hours more and made greater progress toward personal goals than the other participants.

Emmons expanded his research to explore the impact of gratitude on adults suffering from neuromuscular disease. After participants had completed a 21-day gratitude program, researchers sought to determine if there were any physical and socioemotional differences compared to a control group. The gratitude group demonstrated an increase in energy, positive moods, better quality and duration of sleep, as well as a greater sense of overall connectedness and well-being compared to the control group.

All of these gains - positive outlook, physical energy, mental health, motivation toward goals, etc. - have a direct impact on the quality of our relationships. Dr. John Gottman at the University of Washington wanted to determine if there were quantifiable patterns of behavior that would accurately predict the longevity of marriage. Over the past 20 years, he's conducted numerous longitudinal studies and developed a methodology to rate how couples interacted, how positive or negative their expressions were, and how those expressions translated to their spouse. Gottman maintains that he can predict with 90 percent accuracy which marriages will make it and which will end in divorce by observing and calculating their positive and negative expressions when they interact. Unless the couple can maintain a ratio of 5:1 positive to negative, the marriage will more than likely fail. In addition, his formula weights negative expressions and positive expressions differently. According to Gottman, it takes five positive expressions of gratitude, appreciation, or love to counter one negative expression of anger, frustration, or disappointment.

Gratitude Gains in the Workplace

When was the last time you were recognized at work? When was the last time you expressed gratitude to another coworker? Do you remember how you felt? I bet you do. While the personal benefits of gratitude seem obvious, there are professional gains, as well. Implementing gratitude into your organization doesn't have to be an HR initiative rolled out from the leadership team. Whether it is a simple "thank you" note or a more formal expression of appreciation, the psychological effects of gratitude in the workplace can have a tremendous impact on job satisfaction, effort, productivity, and corporate culture. Every single one of us has the power to put gratitude to work – from the custodian to the CEO.

Just like in our personal lives, a sense of gratitude can improve self-esteem, optimism, a sense of unity, and overall well-being at work. Remember the concept of emotional contagion? When we extend expressions of gratitude with our colleagues, we create a "pay it forward" chain of positivity which makes the entire organization more successful. As more people feel the effects, more people will pay it forward. The dopamine effect is a powerful force. If you don't think one person is powerful enough to be a catalyst for positive corporate change, think again.

Tracy Gallimore is the Vice President of Heartland Commerce XPIENT. As a leader in the quick-service and fast-casual markets, the organization has enjoyed steady growth. One of the challenges that often accompanies growth is maintaining a healthy corporate culture. After working with Tracy and his leadership team in a professional development session, I asked them each to identify one or two nuggets that they could put into practice immediately. The guiding mantra of the organization is "everything we do matters," and Tracy recognized that philosophy only works if "everyone matters." So, this concept of gratitude was completely aligned. It wouldn't take a lot of time, and, for a guy who was very much about making lists and checking boxes, this would be a good way to stay focused on the people.

Recently, Tracy shared his experience of writing and delivering that first thank you note. He sketched out a genuine message of appreciation on a little piece of paper and left it for an employee to find on his keyboard when he arrived the next morning. "I felt like Santa Claus!" Tracy told me. "It was just a quick hand-drawn note, but it felt so good to think about him finding it the next morning." And the reaction from the recipient was much different than that from the casual, "Hey, good job!" that we so often give or receive. The employee was grateful - *genuinely grateful* - for Tracy's simple, yet genuine expression of appreciation.

Perhaps what surprised Tracy the most was what has happened after he wrote that first note. First, he was amazed at how contagious something as simple as an expression of gratitude can be. The organization already had a formal employee recognition program in place, and nurturing a healthy culture was an established priority for the company. These little notes were a very simple way to model and reinforce those values by encouraging people to informally recognize one another for contributions, victories, and even challenges. Tracy could have called a meeting with HR and directed them to incorporate the "weekly thank you note" component into the existing program. But the altruistic, organic nature of this new mindset not only reinforced that company initiative from the bottom up, it was a meaningful way for individuals at every level of the organization to take ownership of that culture. Gratitude in the workplace isn't an initiative; it isn't a mandate. It isn't a responsibility defined by title or position. It is an individual and intentional choice to recognize people who matter, and it is contagious!

The second big surprise to Tracy was how being in the market of gratitude - looking for those things for which to be grateful - changed his perspective. "It's easy to focus the problems and the obstacles," he told me. "But, when I know I'm going to write a note that day and I'm looking for the good things, I see things I may have otherwise missed. Sometimes it's hard to decide which good thing to focus on." As it turns out, the virtuous cycle is just as powerful as the vicious cycle, and it makes you aware of some of the good stuff you've been missing.

Dopamine and good vibrations aside, does that actually impact the bottom line of the organization? Francesca Gino at Harvard Business School and Adam Grant of the Wharton School of Business at the University of Pennsylvania explored how being thanked and the perception of being valued affected competence and productivity. In the first experiment, participants were asked to provide feedback on a fictitious cover letter. Half of the subjects received confirmation of their feedback, while the other half received a message that expressed gratitude for completing the task. When the researchers measured the subjects' sense of self-worth afterward, 25% of the group that received confirmation felt higher self-worth compared to 55% of the group that received thanks. The second experiment was an extension of that same group of subjects. Each was asked to provide feedback on another fictitious cover letter. More than double the students in the gratitude group (66%) agreed to provide feedback on the second letter compared to only 32% in the group who received no gratitude.

The "gratitude effect" was explored again in a field study to determine how it impacted productivity. The subjects were fundraisers who all received a fixed salary regardless of the number of calls they made. The director visited one group in person to express appreciation for the job that they did and the contributions they made to the organization. The second group did not receive a visit from the director or expressions of gratitude for their work. That simple demonstration of gratitude generated an increase in the number of calls by more than 50% over the previous week, while the calls of those who had not received thanks remained the same as the previous week.

Still not convinced? An employee survey of more than 2,000 adults conducted by Harris Interactive on behalf of Glassdoor revealed that gratitude is a big driver in employee satisfaction and engagement. A vast majority (81%) of employees reported that they're motivated to work harder when their boss shows appreciation for their work compared to only 37% motivated by the fear of losing their job and 38% motivated by a demanding boss. Moreover, more than half (53%) of employees reported they would stay longer at their company if they felt more appreciation from their boss.

Making Gratitude Stick

It's human nature to get complacent with the things which give us comfort, and it's very easy to take them for granted. Natural disasters like the devastating floods that claimed lives, destroyed entire communities, and cost billions to rebuild in West Virginia and Louisiana this year are a good example. When we hear the news, most of us immediately empathize and sympathize with their plight. We may be grateful that everything we own isn't under water, and perhaps we reach out to help in some way. But, how quickly we forget! That gratitude often fades away as we get on with our lives until ... *poof!* ... it's gone. A few days later when you're complaining about the rain after you've just washed the car or that the weather has ruined your family picnic you've completely forgotten the images of the cars submerged underwater and the despair on the faces of flood victims.

There is an old saying, "If you've forgotten the language of gratitude, you'll never be on speaking terms with happiness." Beyond happiness, we now know that gratitude can be a powerful catalyst for personal and professional growth, mental and physical health, and an overall sense of well-being if we incorporate it into our daily lives with deliberate practice. Here are a few simple strategies to make the power of gratitude work for you.

Look for a reason

Look for a reason to say "thank you." It could be someone in the C-suite, the receptionist, the cleaning person or the cook who made your breakfast at the corner diner. Guide your search with the notion that everyone wants to know that what they do matters. If you really want to make an impact, look for the person that may not realize how important he or she is or who may not hear that what he or she does matters.

Write it down

A personal, handwritten note expressing your gratitude for a particular project or simply for the way someone makes the team better often means more to people than a verbal expression. At a time when every communication is delivered digitally, you'll be amazed at how long people keep these kinds of notes. As a bonus, every time they look at that note or even think about it, they'll get a little boost of dopamine, and you will, too.

Avoid the transaction

A sincere thank you is just that… a thank you. Don't dilute it with "quid pro quo" thinking such as, "I'm going to need his support on this project, so I'll compliment him on that one." Gratitude delivered with a transaction in mind will cheapen the message and likely sabotage the intended results. Sure, you could be strategic in recognizing someone that can help you out later, but it will be much more meaningful if it is solely an expression of sincere gratitude with no expectation in return.

Make it a part of your routine

End each week with a pen and a note card. Think about the positive people in your organization, neighborhood, or community. Think about the people who will take your gratitude and extend it to their work and their colleagues. Choose one person, and end your work week with a sincere expression of gratitude. There is *someone* who is worth 10 minutes of your time, don't you think?

Be grateful for *you*.

Life moves pretty fast, it's really easy to forget about yourself. If you don't take a minute now and again to recognize your own strengths and gifts, you're overlooking the best parts of *you*! Take a moment every single day to identify something good about yourself – not in a narcissistic way – but in a grateful way. While you're brushing your teeth, enjoying that first cup of coffee, or on the commute home from work, give yourself a few minutes each day to recognize progress you've made toward a goal or one way you've helped someone else. Appreciate the good things you've given to the world!

Reflect and Apply

Identify two key nuggets you learned from this chapter.

1.

2.

How can you initiate the "gratitude effect" in your own sphere of influence? Identify two people who deserve a sincere expression of gratitude and a specific action plan for each.

1.

2.

Chapter Nine
****
Happiness is the Secret to Success

The Kingdom of Bhutan is home to approximately 750,000 people in the Eastern Himalayas. In 1972, the fourth King of Bhutan, Jigme Singye Wangchuck introduced a new measurement of national prosperity based on Buddhist spiritual values rather than the gross domestic product. Focusing on the "happiness factor" as a socioeconomic framework, "gross national happiness" (GNH) has since inspired a modern political movement placing happiness on the global agenda. The idea is that sustainable development is only possible if metrics of progress holistically include non-economic aspects of wellbeing. Health. Happiness. A good life.

The rejection of GDP as the only way to measure progress has introduced a new approach to economic development founded on the premise that understanding what humans need to be happy is vital to societal growth. GNH measures fulfilling conditions of "a good life" in 9 domains such as education, leisure time, a sense of belonging in the community, resilience, and how integrated one feels with the culture. In essence, according to the Bhutans, happiness fuels success.

Success is not the key to happiness.
Happiness is the key to success.

Statistics also show that among Bhutanese graduates studying abroad, a high percentage return home to begin their careers even though salaries are significantly less than opportunities overseas. Is it possible that the youngest members of their workforce are choosing happiness over money? The fundamental belief is that people will work harder and contribute more to their organizations and the greater good of society when they are genuinely happy. Distilled down into the simplest terms: true happiness creates sustainable progress because it comes from contributing to the greater good and realizing the brilliant nature of our minds.

It's no secret that we tend to excel at the things that we enjoy or make us happy. It's not a hard and fast rule, but generally, if you love tennis, for example, you'll play more. The more you play, the better you play. And, whatever the "game" - golf, tennis, cooking, political discourse, nuclear physics... pick your passion... we're social creatures, and we tend to "play" more with people we like who share our passion. People who love what they do and love the people they do it with will not only be more engaged, but they are more likely to do it better. Apply the "gross happiness factor" concept to the workplace; everyone impacts organizational happiness, and that metric significantly impacts organizational success. If you subscribe to that, part of your job description is to be happy.

When people are happy, engaged, and contributing, the team is successful. When the team is successful, the members are happier, more engaged, and more willing to contribute. From an organizational standpoint, that's the money shot. But from an individual point of view, that whole construct tends to get distilled down to this:

The harder I work, the more successful I'll be.
The more successful I am, the happier I'll be.

It's easy to get sucked into this logic. The problem is that it's scientifically twisted. There are more than two decades of research that explains what happens in the brain when we experience happiness. Not only do we feel better when we're happy, our brain chemistry physically changes. When we are positive, the brain becomes more engaged, creative, and productive, and that brain activity increases motivation, creative thought, energy, and resilience. Remember all of those happy chemicals? Dopamine, serotonin, oxytocin, and endorphins are released when we experience positive emotions. When they are flowing, the brain works better.

In fact, studies show that we're much more likely to be successful at work when we're happy and positive. Psychology expert and author, Shawn Achor calls this the *happiness advantage*. His research shows a direct correlation between a positive mindset and professional success. Among his findings, sales of happy people are 37% higher, productivity is 31% better, and they are 40% more likely to receive a promotion.

If we know that our level of happiness determines our level of success, why aren't people just happier? Psychologists like Sonja Lyubomirsky at the Greater Good Science Center have been studying happiness to find out just how much control we have over it. Lyubomirsky maintains that one's level of happiness is contingent upon three primary factors: genetic predisposition, external events, and intentional activities.

According to Lyubomirsky, only 10% of our happiness is defined by things we have no control over (losing a job, a car accident, a death in the family) and 50% is attributed to our genetic predisposition for happiness. The remaining 40% is defined by what she calls intentional activity (our behaviors, thoughts, and attitudes). Even the 50% genetic baseline is just a predisposition to happiness, meaning that you can rewire your brain to change it. And that intentional activity which defines 40% of our happiness is entirely within our control.

Break that intentional activity down even further, and it is primarily shaped by three factors:

1. your level of optimism (you have control over things that matter)
2. your social connectedness (positive interactions with others)
3. your perception of stress (challenges, fear, anxieties, or threats)

In other words, your happiness and success largely depend upon how you see yourself and the world. Good news... you have complete control over that!

Consider the following case study on the impact of optimism on success. In the mid-1980s, Metropolitan Life was hiring an average of 5,000 salespeople each year and spending more than $30K each to train them over the course of two years. Of all the new hires, half quit the first year and four out of five within four years. Do the math; the ROI was terrible. At about the same time, psychologist, Dr. Martin Seligman introduced a new theory called *learned optimism*. His theory was that when optimists fail, they attribute the failure to something they can change, not to a weakness or factor out of their control. This difference in attitude, Seligman maintained, enabled optimists to be much more successful than pessimists.

MetLife hired Dr. Seligman to test his theory on 15,000 new MetLife sales consultants. Each of the new hires took two tests. One was the company's regular screening exam, and the other was Dr. Seligman's profile to measure how optimistic they were. He was able to identify one segment of the new hires that failed the company test but scored as "super-optimists" on his profile test. The "super-optimists" outsold the pessimists who passed the company test by 21% in the first year and 57% in the second.

The pessimist sees the difficulties in every opportunity. The optimist sees the opportunity in every difficulty.

Think about how you feel when you see negative news stories in the media - things which you have no control over - such as terrorist attacks, senseless murders, or a plummeting stock market. If you think that doesn't impact your personal happiness and productivity, think again. Seligman continued his research to study the long-term impact of negative and positive news stories on the public, this time with The Huffington Post. Participants were randomly placed into two groups, and each group was shown televised news stories. One group watched 3 minutes of negative news stories. The second group watched 3 minutes of solution-focused news stories of resilience, courage, and accomplishment. It's important to note that these weren't just fun stories of giggling babies or cats watching birds. These were gritty stories about people who persevered and triumphed. For example, one story reported on inner city kids working hard to win a school competition. Another was a 70-year old man who finally passed his GED after failing numerous times. Six hours later, participants were asked to complete a survey that included questions to measure their mood and stress level. The people who watched just three minutes of negative news in the morning were 27% more likely to report having a bad day 6-8 hours later. Conversely, people who watched the same amount of solution-focused stories were 88%

more likely to report having a good day. Positivity and negativity both have sticking power.

Remember the bad chemicals? Visualize what happens to the person who sees the negative world view. His brain is so busy producing all of the stress-fighting chemicals; there is very little bandwidth left to think, process information, be creative, or even recognize the positive events that produce dopamine, oxytocin, and serotonin. The longer one remains in that vicious cycle, the harder it is to get out of it because of the overproduction of cortisol. Eventually, it snowballs to the point where anything good seems completely out of reach, and the "why bother?" attitude takes over. Psychologists call this *learned helplessness*, and there is a direct correlation with learned helplessness, chronic failure, and depression.

The Negativity Bias

The fully developed brain has a built-in negativity bias that is designed to help us survive. It's the protective shield in the brain that is constantly on the lookout for danger or trouble. This is what separates adults with executive functioning skills from fearless toddlers.

This construct is demonstrated in studies that show we recognize angry faces faster than we recognize happy faces. It also explains how people can dislike someone immediately, and it takes a little longer to determine that we like someone. Even when we're relaxed and happy, a part of the brain is always on high-alert for danger or disappointment that could be right around the bend.

The go-to guy on happiness research is psychologist, Daniel Kahneman who received the Nobel Prize for his work in behavior economics. Kahneman has describes what he calls *cognitive traps* between what we remember and what we experience. He claims that because experience and memory are fundamentally different, happiness is impossible to define and remember accurately. How happily we live our lives and how happy our memories of life are may be two very different things. Kahneman found that most people naturally go to greater lengths to protect themselves from a bad experience than they do to experience something good. This is because the brain tends to respond more intensely to negative things than to equally intense positive things.

Imagine you just returned from a 10-day cruise in the Adriatic Sea. The weather was great, the excursions were fantastic, and the Italian delicacies were delicious. However, one bad mussel didn't agree with you and you spent the rest of that day in bed. While the other nine days of the vacation may have been magical, that single negative experience is likely to create a much more powerful memory. Years later, you may only vaguely remember the picturesque drive along the Amalfi Coast or the homemade Limoncello in Sorrento, but the day you were puking up dry toast in your cabin will seem like yesterday. This is all part of the negativity bias in our implicit memory that shapes the way we perceive experiences.

Kahneman also explores the concept of time regarding our perceptions. He maintains that the psychological present is 3 seconds. If you do the math, there are about 25,000 (give or take a few thousand for sleep time) of them in a day. How many of them do you remember at the end of the day? Unless they are intense or surprising, many of the good things leave no imprint on the brain. Unless we train ourselves to recognize them with intentional focus, they simply vanish from our memories.

Conversely, the negativity bias fast-tracks the bad things into memory. It's like a built-in security system that says, "remember this bad thing so you can be on the lookout for it next time." The brain doesn't automatically remember the good things so that you'll notice them next time. You have to consciously pay attention to the good things in order for them to make an imprint on your brain.

The good news is that you can rewire the brain to overcome the negativity bias and take control of the brain makeup that isn't defined by your personality or genetics. Here are four ways to shape those intentional activities and take control of your happiness, nurture your optimism, and experience greater rewards.

Project positivity

Remember that mental states become neural traits when they are intense, prolonged or repeated. You can train your brain to see and experience positive things and create the conditions for the happy chemicals as soon as you wake up in the morning.

Before you put your feet on the floor, take 5 minutes before you're fully awake to think about what lies ahead. Visualize yourself going through your day. Imagine yourself killing that presentation or contributing to a project. Create an "internal forecast" of at least one positive exchange. Visualizing a positive day while in alpha state will produce enough dopamine and serotonin to have a significant impact, and plant some valuable seeds for a positive mindset throughout the day.

Practice positivity

Studies show that when we demonstrate positivity through kindness, the brain produces happy chemicals that make us feel more optimistic and in control of our lives compared to the learned helplessness that comes from an overproduction of stress-fighting chemicals. Random acts of kindness, for example, releases oxytocin. In addition to making us feel more socially connected and loved, oxytocin actually reduces stress and lowers blood pressure.

When you make positive acts an intentional practice, you train your brain to focus on positive interactions rather than negative ones. Simple things like sharing positive praise or expressing sincere appreciation to a colleague in the form of an email will not only give someone else a boost of serotonin but will give you an oxytocin boost, as well.

If you want an extra shot of happy chemicals, send a handwritten note instead of an email. Email is the norm these days, and the surprise of a handwritten note will intensify the positive emotions of the receiver simply because it's unexpected.

Extend positivity

When you notice something good, how often do you take the time to make it a positive experience? How often do you extend your attention to that good thing for more than 3 seconds? It could be something as simple as a compliment from a colleague or the warmth of sunshine when you leave the office.

If you want to convert a good thing into an experience that actually shapes your neural structure, you have to extend the psychological present beyond 3 seconds to create a neural imprint. With deliberation and intentionality, embrace that good thing for 10 seconds, 20 seconds or longer. Shift your focus back to that good thing throughout the day. Each time you do, you produce more positive chemicals that keep your brain humming, and you train your brain to look for the glass half full.

Reflect on positivity

You can't be truly grateful and truly angry or fearful at the same time. The brain just doesn't work that way. Gratitude journals are the simplest way to beat the negativity bias. When you begin and end each day with an intentional focus on those things for which you're grateful, you begin a positive repeated cycle that, over time, will change the physical structure and chemistry of your brain. In addition, just reflecting upon an accomplishment or the belonging you feel when you are part of a team is enough to give you another little boost of oxytocin or dopamine the same way you got a surge when you experienced it.

Understanding the what happens in the brain when we experience certain emotions is just part of the process. Understanding that we can control the neural activity that creates the brain conditions conducive to optimism and happiness is the key.

If we measure happiness by success, we'll never get there. Envision the mechanical rabbit at the racetrack. The dogs can never catch it. It's always just out of reach. Every time you experience success, your own definition of success changes. While getting the VP promotion may be your definition of success today, once you get it, the new measure of success changes to becoming CEO. When you become the CEO, you raise the bar to becoming CEO of a bigger company. If your level of success defines your level of happiness, it will always be just out of reach, just like the mechanical rabbit. One word of caution: there is a very fine line between contentment and complacency.

Happiness isn't just being content with the life we have. It's the optimistic belief that we have control over making our lives better.

Reflect and Apply

Identify one nugget from this chapter that resonates with you. Why do you think it is so meaningful to you?

Describe a time when you either experienced or witnessed the concept of "negativity bias."

List two specific things you can you do to increase the "gross national happiness" of your organization, family or community.

1.

2.

"I'd rather be an optimist and a fool than a pessimist and right."

Part Two

"We act as though comfort and luxury were the chief requirements of life. All that we need to make us happy is something to be enthusiastic about."

Chapter Ten
****
The Art and Science of Journaling

Bert and John were the youngest of six children. Life wasn't always easy for this lower-middle class family struggling to make ends meet. When the boys were in elementary school, their parents were involved in a serious car accident that would injure them both and create significant emotional and financial challenges for the family. While their mother suffered a few broken bones, their father wasn't as lucky. His injuries were more severe causing him to lose total use of his right hand. Unable to work and faced with a long, painful road of physical therapy, he became angry and depressed. The bills began to pile up and the adult stress trickled all the way down the family to the youngest kids. It's safe to say that life was neither perfect nor easy.

Despite the hardships, their mother believed in the power of optimism. Rather than focusing on the challenges, she embraced the opportunity to teach the children a valuable lesson that would stay with them for years. Each night at dinner, she would ask all six kids to share something good about their day. One simple phrase – "tell me something good" – changed the energy of the family and their outlook on life for years to come. "She showed us that optimism is a courageous choice you can make every day, especially in the face of adversity. That optimism was something that our family always had, even when we had little else," the boys would later write in their 2015 book *Life is Good: The Book*.

Today, John and Bert Jacobs will tell you that their mother's optimistic outlook is what inspired them to become entrepreneurs selling T-shirts out of a van at street fairs. After five years of struggling at street fairs and college events, they decided to give it one last shot. They only had enough money to make 48 shirts. John and Bert printed them all with the same design: a smiley face and the words, "Life is good." Within 45 minutes, they had sold every last one. That little company has grown into the $100 million clothing empire, *Life is good*. Far from those difficult times around that cramped kitchen table in Needham, the Jacobs brothers continue to pass along their mother's wisdom to their employees. The simple phrase, "Tell me something good," is one often shared within the organization as a commitment to focus on positivity, creativity, progress, and successes across the organization.

Find the Good Stuff in Every Day

Now that you have a basic understanding of how the brain works and those factors which impact your productivity, creativity, motivation, and overall happiness and mental health, it's time to put them into daily practice with focused intention. The following journal pages are created to provide you with some structure, but it's up to you to apply what you now know about brain function.

Notice that the pages are formatted with free-flowing textures, natural images, and soft colors. In contrast with the way we process the never-ending list of things on our daily to-do list, abstract art is free from the functional restrictions imposed on the visual system and is an effective way to break free from mental roadblocks and engage the whole brain. Recent research in an emerging field called neuroaesthetics has explored the integration of art with whole-brain thinking.

Neuroaesthetics

Founded in 2002, neuroaesthetics is defined as "the scientific study of aesthetic experiences at the neurological level." While aesthetics broadly refers to the process of creating or perceiving art, neuroaesthetics examines what happens in the brain during that process. A large part of the research has focused on the visual qualities that humans find universally appealing (e.g., those related to color, form, or spatial arrangement) and investigate the brain activity and emotional experiences associated with those aesthetic preferences. While this emerging field has been met with criticism from the broader scientific community, the concept of bridging brain science and the visual arts has grown over the last decade with increasing international interest. In a 2014 study published by Vered Aviv in Frontiers in Human Neuroscience, one very specific question was explored: What does abstract art do to the viewer's mind? Aviv's findings indicate that while realistic art activated very specific brain regions, abstract art evoked responses in numerous parts of the brain. Because abstract images do not consist of clearly recognizable objects that the brain tries to identify via memory and association systems, they are processed via brain routes of style and color analysis – pathways less familiar and less utilized. More simply, the brain tends to process abstract images with the whole brain because they just don't fit into one specific region. Free flowing shapes, colors, and lines, therefore, open the mind to think in a less restrictive manner enabling us to form new associations, activate more positive emotions, and potentially form new creative pathways.

We also know that color and mood are inextricably linked. We have innate reactions to colors, and they have psychological properties that relate to the body, mind, emotions and the essential balance between all three. For example, red is a stimulating color that increases your heart rate and grabs your attention. Red screams, "Stop! Pay attention!" Blue is on the other end of the psychological spectrum. The color blue does lower blood pressure but affects us more mentally. Softer blues calm the mind while stronger blues may stimulate clear thought.

The energy we feel from color is both dynamic and personal, and it varies depending on culture, personality, and even the circumstances of the moment. The following list explains the science behind colors, but pay attention to the emotions that they evoke in you. While the soft, free-flowing images on the journal pages are blended washes of colors, the psychological properties may help you harness deeper introspection as you think about each day.

Red: physical
Positive: power, excitement, passion.
Negative: aggression, demanding, danger
Red is the longest wavelength and has the property of appearing closer than it is. This is why it grabs our attention. It is also the most physically stimulating color on the spectrum.

Blue: intellectual
Positive: serenity, reflective, tranquil
Negative: cold, aloof, unfriendly
Blue is mentally calming providing the same psychological affects as the sky or the ocean.

Green: balance
Positive: growth, renewal, natural
Negative: boring, bland, stagnant
Green has the calming effects of nature and the environment. It is the color in the center of the color spectrum reflecting the properties of balance and harmony.

Violet: spiritual
Positive: self-awareness, truth, contemplative
Negative: decadence, superiority, luxury
Violet has the shortest wavelength and encourages deep contemplation and introspection.

Yellow: emotional
Positive: confident, creative, optimistic
Negative: irrational, fear, anxiety
Yellow is stimulating in an emotional sense making it the strongest color psychologically.

Pink: soothing
Positive: tranquil, warm, feminine
Negative: weak, inhibited, emasculating
Pink affects us physically, but in a soothing rather than a stimulating way.

Orange: enthusiasm
Positive: comforting, creative, fun, abundant
Negative: immature, frivolous, silly
Orange is a combination of red and yellow and focuses our minds on physical comfort such as food, warmth, security.

Your Brain on Nature

You'll also notice that the journal pages include watercolor flowers, leaves, and other natural images. Before you dismiss them as "girlie" or "fluffy," the positive effects of nature on mood and stress and mental functioning are well-established. Studies show that regardless of age, gender, or culture, human beings find nature pleasing and relaxing. Moreover, being in nature, or even just viewing natural scenes or objects, reduces anger, fear, anxiety, and stress and is shown to increase a pleasant sense of calmness.

In what has become a classic study of nature on people recovering from gallbladder surgery, Robert Ulrich discovered that patients whose rooms had a view of a green courtyard with plants and trees recovered significantly faster than those patients whose rooms had a view of a brick wall. The patients who saw the natural setting got out of the hospital faster, had fewer complications and required less pain medication than those who had a view of the manmade brick wall. When he extended his research, Ulrich found that simply viewing representations of nature produced similar results. For example, he found that heart surgery patients in intensive care units who were exposed to pictures of natural settings such as trees and water required less pain medication and were less anxious about recovery than patients without access to natural images. Scientists have been studying the effects of nature on mental and physical well-being ever since. We've learned that exposure to nature or natural images reduces blood pressure, heart rate, muscle tension, and the production of cortisol as they increase overall feelings of well-being and positive emotions. Even a single live plant can have a significant impact on stress and anxiety.

Nettie Weinstein expanded upon this body of research in 2009 when she and her colleagues sought to determine whether exposure to nature has an impact on our intrinsic and extrinsic aspirations (e.g., nurturing relationships and making contributions to the community or the greater good vs. making more money, getting promoted, or becoming famous). Participants were randomly assigned to two groups and shown slides of natural images or man-made scenes such as cityscapes. Their findings revealed that, on average, participants who viewed slides with natural scenes had higher intrinsic aspirations that participants who viewed slides with man-made scenes.

Furthermore, they found that participants who viewed slides with natural scenes also had lower extrinsic aspirations that participants who viewed slides with man-made scenes.

Nature restores mental functioning in the same way that food and water refuel our bodies. We spend the majority of our working lives in the manmade world – fighting traffic, racing through airports, staring at computers, cell phones, televisions, spreadsheets, PowerPoints, briefs, proposals... the list goes on. Even simple pleasures such as reading a juicy novel are now delivered on an electronic device. Much like the man-made brick view outside of those hospital rooms, the business and "busy-ness" of everyday life is depleting. Nature and natural images, on the other hand, enable us to replenish exhausted mental resources as they relax the mind and body.

We're often so wrapped up in checking boxes and getting things done; we rarely take the time to sit and reflect. Introspection and reflection require getting to your core, uncovering your values and your obstacles and then deciding the best path to get where you want to be. And if natural images of flowers and leaves can help you get there, why not?

Putting it All to Work

Now, let's put the science of color, nature, and abstract art to work and engage your whole brain in the practice of gratitude. Each spread in the journal has a morning page for you to focus on the day ahead and an evening page for reflection. The pages have been designed with color, nature, and abstract free-flowing images to open the neural pathways in your brain and enhance your introspection as you project and reflect upon each day. Apply what you've learned about happiness, laughter, stress, movement, surprise, and gratitude and the way they all impact the brain with intentionality for the next 28 days.

Take just a few minutes each morning to visualize what the day has in store for you. What will you need to embrace and overcome any anticipated challenges? How can you keep the happy chemicals flowing, and your brain operating effectively? What little steps can you take to get you closer to your personal or professional goals? Before the day gets crazy, think about the people you'll encounter, your personal and professional goals, the attitude you'd like to have, the mark you'd like to leave on the world. Jot down what you need to do to be able to call the day a success.

Before you close the book and start your day, look ahead at the gratitude page. How many things will you need to look for throughout the day? If you know you'll need to list two things for which you're grateful, you'll be more focused on looking for them throughout the day. Finding a certain number of good things is a great way to plug it into your brain as a "relevant task." Knowing that you need to find three things today keeps your brain focusing on which three things are the ones you want to record.

At the end of each day, close your eyes, take a deep breath and tap into your grateful brain. First, look within yourself and be grateful for one thing that you've done to make the world a better place. Finally, close your day with those things "out there" for which you're grateful. You may reflect upon people who have enriched your life or simple things that brought you pleasure like hearing a child laugh or seeing a rainbow. Journaling about the positive things you experience allows your brain to relive them. Also, when we translate an experience into language, we assimilate it into our mental model of the world and create a bolder imprint on the brain.

If you allocate just a few minutes each morning and at the end of your day for introspection and commit to making this part of your daily routine for 28 days, you'll be amazed at the results. At the end of the month, take a look back at your entries. How did they change? How did they change you or the way you see the world?

Enjoy the journey!

"Not everything that can be counted counts, and not everything that counts can be counted."

focus with intention

Date

Today will be a success if I am able to

Three ways to make it happen

surprise can intensify emotions by up to 400%....

gratitude
from the inside out

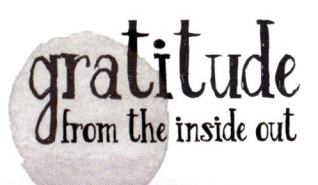

One way I
made the world
a better place today

Two things for
which I'm grateful

focus
with intention

Date

Today will be a success if I am able to

Two ways to get there

the pessimist sees difficulties... the optimist sees opportunities

gratitude
from the inside out

One way I made the world a better place today

Two things for which I'm grateful

focus
with intention

Date

Today will be a success if I am able to

Top priorities:

a lack of creativity is not a lack of imagination.
it's a commitment to the prior art

gratitude
from the inside out

One way I made the world a better place today

Three things for which I'm grateful

focus
with intention

Date

Today will be a success if I am able to

Two goals to make it happen

you did not wake up today to be mediocre...
make the choice to make a difference

gratitude
from the inside out

One way I made the world a better place today

Three things for which I'm grateful

focus
with intention

Date

Today will be a success if I am able to

Top priority for today

live on purpose

gratitude
from the inside out

One way I made the world
a better place today

Three things for which
I'm grateful

focus
with intention

Date

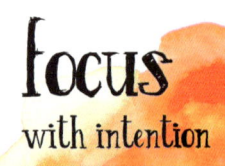

Today will be a success if I am able to

Three ways to get there

the brain cannot make its own fuel...we have to create it

gratitude
from the inside out

One way I made the world
a better place today

Three things for which I'm
grateful

focus
with intention

Date

Today will be a success if I am able to

My top priority

today is a gift... no exchanges, no returns.

gratitude
from the inside out

One way I made the world a better place today

Today made me realize how grateful I am for....

… # focus
with intention

Date

Today will be a success if I am able to

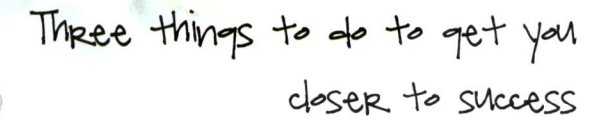

Three things to do to get you closer to success

laughter is contagious… so is stress

gratitude
from the inside out

One way I made the world
a better place today

Three things
for which
I'm grateful

focus
with intention

Date

Today will be a success if I am able to

Top priorities

there are seven days in a week…..."someday" isn't one of them.

gratitude
from the inside out

One way I made the world a better place today

Three things for which I'm grateful

focus
with intention

Date

Today will be a success if I am able to

Three goals to get there

here. this. now… be present.

gratitude
from the inside out

One way I
made the world
a better place today

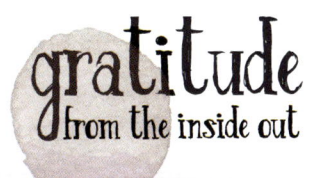

Two things for
which I'm grateful

focus
with intention

Date

Today will be a success if I am able to

Two things I can do today to get me closer to where I want to be

surprise yourself. be unpredictable. be open to discovery

gratitude
from the inside out

One way I made the world a better place today

Three things for which I'm grateful

focus
with intention

Date

Today will be a success if I am able to

Two things I can do to get me closer to where I want to be

people who smile appear more trustworthy and confident

gratitude
from the inside out

One way I made the world a better place today

Three things for which I'm grateful

focus
with intention

Date

Today will be a success if I am able to

Two things I can do today to get me closer to my goal

don't fear the unknown...learn how to embrace it as an opportunity to discover

gratitude
from the inside out

One way I made the world a better place today

Two things for which I'm grateful

focus
with intention

Date

Today will be a success if I am able to

Two things that will get me closer to my goals

train your brain to look for the good stuff

gratitude
from the inside out

One way I made the world
a better place today

Three things for which I'm
grateful

focus
with intention

Date

Today will be a success if I am able to

2 ways to make it happen

a hearty belly laugh boosts the immune system and decreases stress

gratitude
from the inside out

One way I made the world a better place today

Two things for which I'm grateful

focus
with intention

Date

Today will be a success if I am able to

the best ways to make that happen

project positivity...

gratitude
from the inside out

One way I made the world a better place today

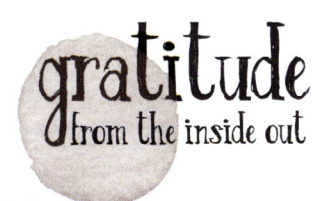

Four things for which I'm grateful

focus
with intention

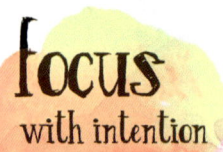

Date

Today will be a success if I am able to

Three specific ways to make it happen

give yourself permission to be vulnerable... be open to new possibilities

gratitude
from the inside out

One way I made the
world a better place today

Three things for
which I'm
grateful

focus
with intention

Date

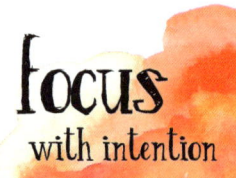

Today will be a success if I am able to

Top priorities

live the life you love; love the life you live...

gratitude
from the inside out

One way I made the world
a better place today

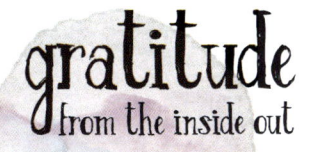

Two things for which I'm
grateful

focus
with intention

Date

Today will be a success if I am able to

Two goals to get there...

look for reasons to say "thank you"...

gratitude
from the inside out

One way I made the world a better place today

Three things for which I'm grateful

focus
with intention

Date

Today will be a success if I am able to

Top priorities

the secret to life is learning how to enjoy the ride....

gratitude
from the inside out

One way I made the world a better place today

Three things for which I'm grateful

focus
with intention

Date

Today will be a success if I am able to

Today's priorities

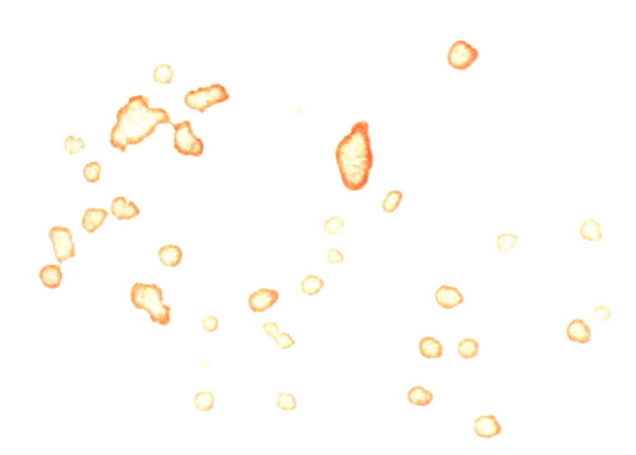

expressing thoughts in 'free form' mode unlocks imagination and creativity...

gratitude
from the inside out

One way I made the world a better place today

Three things for which I'm grateful

focus
with intention

Date

Today will be a success if I am able to

Three things I must do to make that happen

surprise can intensify emotions by up to 400%....

gratitude
from the inside out

One way I made the world a better place today

Today made me realize how grateful I am for....

focus
with intention

Date

Today will be a success if I am able to

Two ways to make it happen

○

○

smiling reduces the heart rate, blood pressure, and cortisol levels...

gratitude
from the inside out

One way I made the world
a better place today

Three things for
which I'm grateful

focus
with intention

Date

Today will be a success if I am able to

Top priorities:

the ability to learn is a skill; the desire to learn is a choice...

gratitude
from the inside out

One way I made the world a better place today

Three things for which I'm grateful

focus
with intention

Date

Today will be a success if I am able to

Three goals to make it happen

happiness is the key to success...

gratitude
from the inside out

One way I made the world a better place today

Three things for which I'm grateful

focus
with intention

Date

Today will be a success if I am able to

Top top priorities

move the body and the brain will follow....

gratitude
from the inside out

One way I made the world a better place today

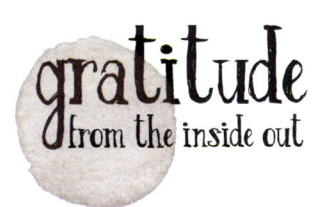

Two things for which I'm grateful

focus
with intention

Date

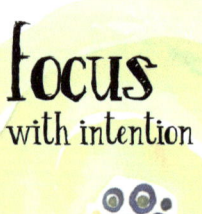

Today will be a success if I am able to

Top priorities

Leave a mark, not a scar....

gratitude
from the inside out

One way I made the world
a better place today

Two things for
which I'm grateful

focus
with intention

Date

Today will be a success if I am able to

Two goals to get there

laughter releases oxytocin and strengthens the human bond ....

gratitude
from the inside out

One way I made the world a better place today

Three things for which I'm grateful

"Few are those who see with their own eyes and feel with their own hearts."

A few final notes of gratitude...

To Mom and Dad,
You recognized early on that I march to the beat of a very different drummer, and you made extraordinary sacrifices so that drum beat never faded. You instilled the value of setting high goals and the power of developing the work ethic and grit necessary to reach those goals.

To Mrs. Alice Austin,
You taught me that our ideas are only as good as our ability to communicate them. You filled my cognitive backpack with invaluable skills that have served me well throughout my lifetime, and your mastery of language is something to which I continue to aspire.

To Christopher,
I am fortunate and grateful for your unconditional love, support, and encouragement. You inspire me to climb mountains that I may not otherwise have climbed. Thank you for believing in me, learning with me, and laughing with me. Most of all, thank you for letting me be me.

References

Achor, Shawn. *The Happiness Advantage: The Seven Principles That Fuel Success and Performance at Work,* 2011.

Alfini, Alfonso, et. al. Hippocampal and cerebral blood flow after exercise cessation in master athletes. Frontiers in Aging Neuroscience, August, 2016, http://dx.doi.org/10.3389/fnagi.2016.00184

Aviv, Vered. What does the brain tell us about abstract art? Frontiers in Human Neuroscience, February, 2014.

Bender, Rachel Grumman. How Color Affects Mood, Huffington Post, 2013.

Berk, Lee and Tan, Stanley A. *Cortisol and Catecholamine Stress Hormone Decrease Is Associated with the Behavior of Perceptual Anticipation of Mirthful Laughter.* The FASEB Journal. 2008;22:946.11.

Brown, Brené http://brenebrown.com

Brown, Sunni. *The Doodle Revolution: Unlock the Power to Think Differently,* 2015.

Centorrino, Samuele; Djemai, Elodie; Hopfensitz, Astrid; Milinski, Manfred; Seabright, Paul. Honest signaling in trust interactions: smiles rated as genuine induce trust and signal higher earning opportunities. January 2015, Volume 36, Issue 1.

Emmons, Robert. http://emmons.faculty.ucdavis.edu

Fulghum, Robert. *All I Really Need to Know I Learned in Kindergarten,* 2004.

Glenn R. Fox,* Jonas Kaplan, Hanna Damasio, and Antonio Damasio. Neural correlates of gratitude. Frontiers in Psychology. 2015; 6: 1491. Published online 2015 Sep 30.

Grant AM, Gino F. A little thanks goes a long way: Explaining why gratitude expressions motivate prosocial behavior. J Pers Soc Psychol. 2010 Jun;98(6):946-55.

Greenberg, David. *Presidential Doodles: Two Centuries of Scribbles, Squiggles, Scratches & Scrawls from the Oval Office,* 2007.

Goewey, Don Joseph. The Neuroscience Behind the Competitive Edge, 2015.

Gottman, John. https://www.gottman.com/

Humor, Laughter, and Those Aha Moments. *On the Brain*, The Harvard Mahoney Neuroscience Institute, Spring 2010.

Isaacson, Walter. *Einstein,* 2011.

Jacob, Bert and Jacobs, John. *Life is Good, 2015.*

Kahneman, Daniel. *Experienced Utility and Objective Happiness: A Moment-Based Approach* in Choices, Values and Frames. (2000).

Kahneman, Daniel. Th*inking Fast and Slow*, 2011.

Kirby Elizabeth, Daniela Kaufer, et. al., Acute stress enhances adult rat hippocampal neurogenesis and activation of newborn neurons via secreted astrocytic FGF2. Published online 2013 Apr 16.

Leberecht, Tim. In the Age of Loneliness, Connections at Work Matter, *Harvard Business Review*. Sept. 18, 2015.

Luna, Tania and Renninger, Tania. *Surprise: Embrace the Unpredictbale and Engineer the Unexpected,* 2015.

Lyubomirsky, Sonja. The Greater Good Science Center, http://greatergood.berkeley.edu/author/sonja_lyubomirsky

Maguire EA, Woollett K, Spiers HJ. London taxi drivers and bus drivers: a structural MRI and neuropsychological analysis. Hippocampus. 2006;16(12):1091-101.

Paterniti, Michael. *Driving Mr. Albert: A Trip Across America,* 2001.

Pink, Daniel. Drive: *The Surprising Truth About What Motivates Us*, 2009.

"The Real Rain Man", documentary by Focus Productions, Bristol, England, UK, 2006.

NASA studies mega-savant Peek's brain". *USA Today*. Associated Press. 2004.

Roberts, Siobhan "A Hands-On Approach to Studying the Brain, Even Einstein's". *The New York Times*, 14 November, 2006.

Rohde, Mike. *The Sketchnotes Handbook: The Illustrated Guide to Visual Notetaking*, 2012.

Seligman, Martin. *Learned Optimism: How to Change Your Mind and Your Life*, 2006.

Schwartz, Tony. Why Fear Kills Productivity, *New York Times,* December 5, 2014.

Schwartz, Tony. http://theenergyproject.com/

Sinek, Simon. *Start with Why*, 2009.

Sinek, Simon. http://startwithwhy.com

Sukel, Kayt. Neurotransmitters. *The Dana Foundation*, 2012.

Ulrich, Roger. View through a window may influence recovery from surgery. American Association for the Advancement of Science. April 27, 1984.

Weiwei Men, et al., "The corpus callosum of Albert Einstein's brain: another clue to his high intelligence?," *Brain*, 2013.

Witelson, Sandra. "The exceptional brain of Albert Einstein". *Lancet*. Volume 353 (9170): 2149–53.

Today's leading companies need people who know how to
reLEARN®
A SMARTER WAY TO WORK

- ☑ **Increase employee engagement and retention**
- ☑ **Maximize your multi-generational workforce**
- ☑ **Enhance creativity and innovation**
- ☑ **Improve communication**
- ☑ **Outsmart the competition**

In a multi-generational workplace with Millennials who account for nearly half of the employees in the world, identifying those factors which impact productivity and organizational learning has become increasingly more challenging. Understanding how we learn and how different learning styles impact the way people work, solve problems, and communicate can give your company a clear and significant advantage. Collaborative learning, problem solving and engagement are essential to innovation, productivity, and a healthy organizational culture. As more and more companies strive to gain marketshare, the winners will be those that create optimal learning environments and apply brain-based research to the way they work.

Maximize the intellectual capital of your organization with whole-brain thinking and learning.

ANDRICK GROUP™

Learn more about how to give your team a competitive edge with the strategic dimensions of whole-brain thinking and learning.
www.andrickgroup.com